AF560757

PUNJAB PEASANTRY IN TURMOIL

Punjab Peasantry in Turmoil

Edited by
BIRINDER PAL SINGH

MANOHAR
2010

First published 2010

ISBN 978-81-7304-866-1

Published by
Ajay Kumar Jain *for*
Manohar Publishers & Distributors
4753/23 Ansari Road, Daryaganj
New Delhi 110 002

Typeset by
Kohli Print
Delhi 110 051

Printed at
Salasar Imaging Systems
Delhi 110 035

To all
those peasants
who fed the world,
and finally
got fed up
with themselves

Jattan Di

Jattan de khetan vich
suraj ugda hai
phir vi ajj na kall hai
yaro jattan di.
Sare jag nu vasda
karke ki khatya
chal-chala-chal-chal hai
yaro jattan di.

Of Jutts

From the fields of Jutts
Rises the sun
But O my friends
Today nor tomorrow
Belongs to them.
What do they gain
Sustaining the whole world
When my friend
They are ever in turmoil
Ever in doldrums

PASH

Contents

Introduction

Birinder Pal Singh

A peasant economy, according to Daniel Thorner, can be characterized on the basis of five criteria, namely, (i) half the population must be agricultural, (ii) more than half of the working population must be engaged in agriculture, (iii) existence of a 'territorial State', (iv) a division of society between towns and countryside and the most important of all, (v) the representative units of production being peasant family households (Thorner, 1971). Following this definition, most of the world's population barring North America, Europe and Australia—which incidentally have low population density—may still be called to constitute a peasant economy. And agriculture has remained the dominant mode of subsistence ever since humankind adopted a settled mode of living.

The peasant society is a logical consequence of the peasant economy that may still be found in the larger parts of the world. For Robert Redfield: 'Peasant society and culture has something generic about it. It is a kind of arrangement of humanity with some similarities all over the world.' It is a 'type without localization' (Shanin, 1971: 240). Providing an analytical definition of peasantry, Shanin says:

> The peasantry consists of small agricultural producers who, with the help of simple equipment and the labour of their families, produce mainly for their own consumption and for the fulfilment of obligations to the holders of political and economic power. (ibid.: 240)

This definition suggests that peasantry is necessarily tied to a landholding (predominantly small), a family, virtually fixed in space, hence, a more or less permanent neighbourhood, giving rise to a community and a we-feeling that Emile Durkheim calls mechanical solidarity. The stability and prolonged existence of such a social fabric has given rise to norms and values in which the collective conscience gives primacy to the interests of the group rather than the individual. This has given rise to a culture, a *weltanschauung*, a cosmology characteristic of a peasant society that is primarily collectivity or community-oriented. The strength and tenacity of this social structure was such that it remained impervious to even strong forces of the outside world. Karl Marx mentions that in rural societies the operation of 'law of value' never emerged. Therefore, the 'general economic laws of society' did not apply. Thomas and Znaniecki have documented at length the basically social, rather than economic way of reasoning, the lack of calculation (that is of seeking to maximize income in money terms) of the peasantry. They have also indexed the 'irrational' behaviour of peasants as far as land, loans, prices and income, etc., are concerned (ibid.: 246). This characterization of the Polish peasant fits well on the Punjabi peasantry. All of us are aware of the innumerable number of the anecdotes that testify similar traits of the Jutts (Sikh) and Jats (Hindu) as well, of this region.[1]

Punjab, the land of the five rivers, has been known for its agricultural fertility since antiquity. The Punjabis thus have naturally developed skills for cultivating crops. They are hardy people who undertake any kind of physical toil to seek prosperity. The fertile cultivable land made them prosperous and ingrained in them a sense of security that of course is characteristic of all peasantry to some extent. But Punjabi peasantry also combines with it a spirit of endurance and adventure. The geophysical location of Punjab, as the gateway to Hindustan, exposed its people to frequent and ferocious invasions from outsiders, starting from Alexander the Great to Ahmad Shah Abdali. The British who annexed Punjab last of all in 1849, however, did not follow this route. A peasant who is improvident becomes doubly so in Punjab because of the abovementioned reason. He cannot afford to save and store in the wake of frequent invasions and consequent devastation. That is how he has learnt to live in the present and enjoy life to its full. There is a saying in Punjab: *Khada-peeta lahe da, baqi Ahmad Shahe da* (Whatever you consume is yours, the remainder shall be taken away by Ahmad Shah [Abdali]).

In the words of Niharranjan Ray:

> History therefore taught the Panjab and her people one very important lesson, namely, not to forget or be oblivious of temporal or secular situations of any given time or space, howsoever engrossed one might find oneself in matters of the mind and the spirit. (Ray, 1975: 105)

This was further reinforced by the stamping of religious legitimacy granted by Sikhism.

> Guru Nanak was very particular on one aspect of the way of life he had initiated, namely, to take into consideration the socio-temporal, that is, the secular aspect of life with as much seriousness as the ethical and the spiritual. This provided a solid material base for the community of disciples he sought to bring to being. (ibid.: 107)

The above-mentioned factors combined with extreme weather conditions made the Punjabis more hardy and industrious.[2] A settlement officer records in 1910 about the Jat character:

> . . . unremitting in toil, thrifty to the verge of parsimony, self-reliant in adversity, and enterprising in prosperity, the Jat . . . is the ideal cultivator and revenue-payer. Ploughing, weeding or reaping, he will bear the burden and the heat of the day, and at night take his turn at the well. Of the same fibre are the women, and if the Rajput wife is an economic burden, the Jatni is an economic treasure. (Darling, 1977: 35)

The *Jatni* provides help to her husband in all types of agricultural work at all levels, be it harvesting, threshing, and so on but for the sowing. Besides, she performs all the household chores including the caring of milch and other cattle. These chores are no less arduous than the work in the field and she is still a lively being. Her different chores at home and outside, for instance carrying food (*bhata*) for her husband toiling in the fields, make rich material for the folk songs of this region. She is a thrifty, dynamic and robust person. When she 'dances in Patiala its tremors can be felt in Jullundur', says a folk song (*Jad nachchan main Patiale dhamak Jalandhar paindi*).

Darling notes: 'The mass of rural population of the Punjab, however, were neither landlords nor tenants; they were the families of petty proprietors' (ibid.: xviii). This is the legacy of the Sikh rule and the Sikh religious tradition that dictates equality of castes and classes. After the demise of Guru Gobind Singh, Banda Bahadur took the reigns of Sikhism and fought against the rulers to abolish estates and provide land to the tillers. After the fall of Sirhind in May 1710, Banda Bahadur took control of the area from Karnal to Ludhiana. Ganda Singh writes about him:

> He had no time to organize any regular administration. But he knew the woes of the peasantry, groaning under the oppression of the landlords, and took immediate steps to abolish the Zimidari system. This was a revolutionary measure which exercised a great influence on the future fiscal history of the Punjab. (Ganda Singh, 1956: 19)

He continues:

> As a result of the application of his methods the tillers of the soil soon became masters of their holdings and in the course of time the curse of the Zimidari system which till very recently afflicted many parts of India, was lifted from the Punjab. (ibid.: 20)

Subsequently, the most powerful Sikh ruler Ranjit Singh is believed to have completely ousted the *taluqdars* (the holders of a superior right over land, largely the descendants of former chiefs, *jagirdars, ijaradars* and other officials) from certain parts of his dominion (Banga, 1978). Baden Powell notes that the proportion of area held by all the big and small zamindars in the total cultivated area in the Punjab was 10 per cent since 'the Sikh rulers encouraged the actual cultivators as against the holders of superior ownership, which "levelled down the differences and compelled an equality of the landlord and inferior"' (ibid.: 171). On the basis of the *Settlement Reports*, Banga notes:

> By far the bulk of land in the Panjab during the Sikh times was held by small proprietors who cultivated their lands in whole or in part. In Lahore, for instance, over 27 lakh acres of land was cultivated by the proprietors themselves, while the land given to tenants amounted to over 3 lakh acres only. (ibid.: 174)

The Malwa region saw strong movements against the landlords, and absentee landlordism, both before and after the Partition of the country, in the erstwhile east Punjab. After India attained

Independence, land reforms, including land ceiling, were implemented more rigorously in Punjab than in many other states in the country.[3] This continued the legacy of the earlier times of small proprietorship and equality amongst the peasantry. The size of the operational landholdings in the reorganized Punjab remains largely concentrated in the small size, though lately marginal landholdings are disappearing owing to fragmentation of land due to increasing population, the breaking down of the joint family and the tightening grip of the capitalist market. According to the *Agriculture Census, Punjab 1990–1*, the size of operational landholdings of up to four hectares was 76.98 per cent in 1970–1; 66.61 per cent in 1980–1; 69.41 per cent in 1985–6 and 70.58 per cent in 1990–1 (*Statistical Abstract of Punjab*, 2004).

The Sikh religious tradition and administrative check on the enlargement of landholdings together have given rise to a milieu of equality in the rural areas of the present, east, that is, Indian Punjab. The peasants amongst themselves and with their workers from lower castes did not keep distance from each other, characteristic of peasant/rural societies in other parts of India. In her anthropological enquiry of the Jutt Sikhs whom she calls *The Robber Noblemen*, Pettigrew notes: 'The way of life of the Jats [read Jutt] is essentially the same whether they were rich or poor' (Pettigrew, 1975: 38). 'Moreover, labourers, whether Jat or Mazhabi, were treated exactly alike by their Jat employers. The core of the relationship was the same in all these instances and unaffected by caste' (ibid.: 45).

A look into the early history of the Jutts provides the clue. Irfan Habib writes:

> We have seen that in Sind the Jatts [read Jutts] had been a large primitive community based on pastoral economy and with an egalitarian or semi-egalitarian social structure. They had 'neither rich nor poor', nor 'small nor great', in the words of two quite independent accounts. (Habib, 1976: 95)

The historical sources suggest that the pastoral tribe of Sind living on either side of the Indus gradually moved to this land of five rivers. Their expansion took place between the eleventh and sixteenth centuries. They settled in this region at the beginning of the sixteenth century when they encountered the Persian Wheel.

This converted them into cultivators. But socially their status was still low. According to the *Dabistan-i-Mazahib*, Jatts were 'the lowest caste of the Vaishyas even in the middle of the seventeenth century' (ibid.: 97).

Why did the peasantry (Jutts) in Punjab accept Sikhism? Habib suggests: 'the Jatts received Sikhism at a time when by historical circumstances they were in need of it most . . .' (ibid.: 99). More than the problem of Sanskritization was the problem of revenue burden they had to bear in the seventeenth century. This oppression from the ruling classes of the Mughal Empire pushed them towards armed violence. Habib concludes: 'This further cemented the historical association between the Jatt peasantry and Sikhism, though the association certainly antedates the agrarian crisis of the Mughal empire' (ibid.: 100).

The Jutts not only accepted Sikhism but also joined the Sikh army, first organized by the sixth guru Hargobind. This army also consisted of certain anti-social elements and marginal castes and classes as well. Habib argues that the shift from Khatri to Jatt leadership within the Sikh movement had already taken place by the middle of the seventeenth century. He quotes the *Dabistan-i-Mazahib* that although the gurus had been Khatris, 'they have made the Khatris subservient to the Jatts' (ibid.: 98). Since then they have remained in the forefront of every militant movement launched by the Sikhs against their adversaries which included the Afghan invaders, the Mughal emperors, the British imperialists, the Indian capitalism and lately for a confederation or autonomy from the Government of India. They, however, did not undermine other castes. The formation of *misls* (confederacies) in the eighteenth century is an illustration *par excellence* of equality and fraternity. Many chiefs belonged to castes lower than the Jutts but there was no discrimination against them.[4]

This elevation of the lower castes has its roots in the period preceding the *misls*. In the words of Ganda Singh: 'Banda Singh was a great political leveller and a thorough social uplifter. Wherever he went, he raised the down-trodden to position of authority and social prestige' (Singh, 1956: 20). He further quotes William Irvine:

> In all the parganas occupied by the Sikhs, the reversal of the previous customs was striking and complete. A low scavenger or leather dresser, the lowest of the low in Indian estimation had only to leave home and join the Guru [meaning Banda Singh] when in a short space of time he would return to his birth-place as its ruler, with his order of appointment in his hand. As soon as he set foot within the boundaries, the well-born and wealthy went out to greet him and escort him home. (ibid.: 20)

The organization of Dal Khalsa, according to Niharranjan Ray, has been rightly characterized as a landmark in the history of Sikhs. The twelve *jathas* established in different parts of Punjab, which were organized by Nawab Kapur Singh into Dal Khalsa, gradually evolved into territorial powers. These *jathas* came to be known as *misls*, an Arabic word meaning 'equal', since 'all the *jathas* were equal in status and there was complete equality among the members of any given *jatha*' (Ray, 1975: 112). Any Sikh who had a horse of his own had the freedom to join any *jatha* which had its own leader. Each *jatha* could also undertake an expedition or go to war under its own leader, but under the banner of Dal Khalsa. 'All booty collected or acquired had to be shared by all the other *jathas*, each in proportion to its own strength' (ibid.: 111).

According to another version *misl* means an account book. Since the account books of all the twelve *jathas* were lying at the Akal Takht, these *jathas* came to be known as *misls*. All *jathas* assembled here twice a year for stocktaking and laid down general policies and programmes for the Sikhs. This assembly was called the Sarbat Khalsa. According to Ray, the institution of Khalsa or Dal Khalsa was so strong that 'Even a powerful leader like Ranjit Singh did not dare describe himself as *Maharaja* or forget to acknowledge the overall supremacy of and allegiance to the Khalsa' (ibid.: 113).

The sharing of booty and taking a collective decision for the *jatha* or the community have been the hallmarks of the Sikh religion and philosophy. Guru Nanak had emphasized *Kirat Karo* (do labour) and *Wand Chhako* (share with others) meaning thereby that there is no place for non-workers or idlers in Sikhism. It also laid stress on the fact that each one must share the product of his/her labour with fellow individuals, thus instilling in them a feeling of non-possession or non-accumulation of wealth material and strengthening 'we' feeling in the community. It enhanced the social solidarity amongst members. Thus, the community could stand like a rock against any oppressive assault on them. The tenth guru Gobind Singh instilled in the Khalsa, the Sikh fraternity, fearlessness and strength neither to oppress anyone nor tolerate anybody's oppression. The Khalsa is ever at war against injustice, among other things. These objectives could be achieved if there was solidarity in the community. It is this cohesion that was further maintained and reinforced in collective decision-making. The *misls* were putting these Sikh tenets into practice. That is why the eighteenth century is considered the most glorious period in Sikh history from the perspective of theory and practice of the Khalsa ideology.

Shanin argues that peasant thinking often seems to the outsider to be capricious and subjective, containing large elements of what may be called pre-Socratic thought, in which two contradictory opinions may be held simultaneously. The same is also true of the Punjabi Jutts. The conflict between the social value of equality and the political value or orientation of domination and subordination is the result of history's weight on the Jutt mind/psyche. Their egalitarianism of pre-Sikh conversion period was reinforced by the thrust on equality preached by the Sikh religion and philosophy. And the political value of fighting against the tyrant and oppressor, hence against someone who dominates, was further re-invigorated with the creation of the Khalsa. The tenets of Khalsa ideology expects an *amritdhari* (the baptised) to fight against any kind of domination or authority that violates the freedom of any individual let alone himself. A Khalsa is answerable only to *Sacha Padshah*, the God, and no other mundane authority since it pertains to deceit and corruption and hence referred to as *jhutha padshah*.

The social egalitarianism and collective opposition to authority or state power were, however, retained over time while the value of political equality got drained with the waters of history. The loss of this value might be attributed to the supremacy of Maharaja Ranjit Singh (1789–1839) and later of other kings of smaller states, lesser feudatories and sardars of estates who encouraged personal loyalty and sycophancy and promoted class inequality.[5] It seems that the so-called *Sarkar-i-Khalsa* (government of the Khalsa) and

Darbar Khalsaji (the court of the Khalsa) did not consolidate the *misls* to establish the Khalsa *raj* in the real sense of the term, though it was claimed to be so. The state of Maharaja Ranjit Singh was definitely superior to, and might have been more benevolent than the neighbouring and other states, but that does not let it qualify for the Khalsa *raj*.

The period of kings and princely states under the colonial rule generated social and political inequalities that were further strengthened in the post-colonial period with the development of market economy and liberal democracy. The nature of private property, that is, land and *mal*[6] (cattle), in a class society sustained the dichotomy of domination/subordination in Jutt/peasant society. Therefore, fighting against the oppressive authority for the enlargement of space for domination is juxtaposed to subordination of others and establishing *sardari* (supremacy) or *chaudhar* (authority) of oneself. This conflict of values is manifest in all movements and institutions dominated by the Jutt Sikhs.

It is this notion of *sardari* or *chaudhar* that forms the foundation of feeling of revenge and honour, so very crucial to the Jutt character. The logic of murder also flows from it. The chain reaction once started lasts up to seven generations in Jutt society. To have one's moustache straightened up (*muchh khari rakh ke*) and walk with head held high (*sir chakk ke*) with stiffened upright neck (*dhaun akra ke*) are prominent traits of a Jutt Sikh personality. All these elements constitute *sardari* that defies subservience. It is recorded in *Haqiqat-i-Bina* that 'everybody amongst the Sikhs was "independent in his own place"; if any one possessed even two horses and a village he was "not subservient to anyone else" ' (Banga, 1978: 33). Princep also notes that Punjab in the late eighteenth century was 'ruled by seventy thousand sovereigns' (ibid.: 34).

Throughout the eighteenth century the Sikhs had a trying time as the Mughal rulers followed the policy of extermination.

> What kept the spirit and morale of the common Sikh was their faith in the Khalsa and the *Sri Guru Granth Sahib*; the more persecuted were they, the more wretched were the conditions in which they had to live, the more were their determination not to yield, not to give up. (Ray, 1975: 110)

Till the advent of colonialism, the Punjabi peasant relied on nature for agricultural production. The British introduced modern technology, dug canals (the famous canal colonies, now in the west Punjab), placated the Raj loyalists by granting large chunks of land in the new colonies; built dams and generated electricity. These were essential inputs into the later day Green Revolution that took place in east Punjab during the 1960s.

After 1947, the Government of India adopted modern science and technology to usher the country into the new age. The Punjab Agricultural University was established at Ludhiana in 1962, with centres at Hisar (now in Haryana) and Palampur (now in Himachal Pradesh), to promote agriculture and later horticulture, to furbish the godowns of grain to feed the teeming millions in the rest of the country. The Punjabis, true to their temperament, took to modern technology as that would reduce their physical drudgery, enhance crop production manifold and hence increase profit in the modern-day market society.

The surplus production as a result of the Green Revolution made the peasants of eastern Punjab prosperous. They invested not only in agricultural implements but also in consumer goods to seek physical and sensual gratification. Unfortunately, it led to a lopsided development of the Punjabi society. The agricultural development and farmers' prosperity did not uplift the village or the rural area but promoted the growth of urban centres. The *mandi* towns were developed by the state government, ushering urbanization and modernization, rather than containing the benefits of surplus production in the village itself. The Gandhian objective of village-centric development was ignored by the government if not formally shelved (Singh, 2005).

But the industrious Punjab peasantry worked day and night to increase surplus production of every crop it cultivated. A glance at the statistics would show that even during the peak period of Sikh militancy when the farmers preferred to remain indoors after dusk for fear of the militants and the police, the production of wheat and rice did not decline.[7] So much so that over the last few years there has been a problem of storing wheat and rice in the government godowns. The problem of marketing these two major crops of this region, for whatever reasons, still persists and has virtually become an annual feature. This has forced the farmers to strike against the government policies of procurement, thus coercing the ruling party, be it the Congress or the Akali Dal, to impress upon the union government to keep procuring the wheat and rice for the central pool with ensured sufficient rise in the minimum support price (MSP). This of course is political populism or opportunism if one may choose to call it that, which political parties indulge in to placate the rural voter.

Since the late 1960s the Punjab farmer has brought even more land under cultivation,[8] dug more tube wells[9] that are now being replaced by high-powered submersible pumps as the water table is lowering at a rapid speed, almost a metre per year, in most parts of the state. The number of tractors, harvest combines and allied agricultural implements have also risen consistently over the last four decades.[10] The existing canals have been brick-lined as also the newer and numerous of their tributaries, though dependence of the farmer on canal water has not increased over the years.[11]

In the name of developing the villages, schools were opened all over the state. The recently acquired prosperity motivated the small and middle peasants to spare their children for schooling and higher education so that they could become government officers with salary and status. Ever more schools and colleges[12] and the growing list of student enrolment gave a false sense of modernization and development to the state government and other agencies. But the government did not generate avenues for the absorption of the educated, thus generating unemployment that has grown by leaps over the last few decades.[13]

The unplanned and unregulated Green Revolution soon showed signs of malaise. Briefly put, there exists a serious problem of unemployment, especially amongst the youth. It has led to mass exodus of this section of population in search of greener pastures, wherever available, no matter what the cost. They do not mind even playing with their lives. The Malta boat tragedy has not dampened the spirit of Punjabi youth in seeking illegal immigration.[14] The situation has become so bad that not only the poor, the marginal and the unemployed youth, even the well-off ones and above all those with some technical expertise, not to talk of the scientific and technical professionals, are migrating to the West, much to the detriment of the society of their birth and loss to its economy.[15] The peasant families are desperate to marry their sons and daughters abroad in the hope of salvaging and rescuing the whole family subsequently, no matter what may happen to the individual migrating. The growing numbers of failed marriages with the Non-Resident Indians (NRIs) is an issue of serious concern not only for the people but also for the governments of the two countries.[16] There is extensive selling and mortgaging of land, in such pursuits and a mushrooming of travel agencies and marriage bureaus. The poor innocent peasants are getting involved in innumerable frauds and the number of these frauds is rising every day.

The ecology and environment of this region is another victim of the Green Revolution. The forest cover in eastern Punjab hovers around 5.82 per cent which is far below the minimum average of 33 per cent.[17] And this too includes the Shivalik foothills. The cultivation of rice in a hot and dry region with largely sandy porous land has played havoc with the water table. The hybrid seeds of the high yielding varieties (HYVs) of grain are not only water thirsty compared to the indigenous ones, but also highly dependent on fertilizers and pesticides.[18] Soil fertility levels have fallen and pests have become resistant to pesticides, adding to the woes of the marginal and middle farmers.[19] They find it difficult to manage the capital-intensive agriculture. They have to depend on the commission agents for ready cash and loans, thus increasing their debts every season (Shergill, 1998). The rising debts have pushed these peasants towards suicide. This issue is quite antithetical to the very temperament of the hardy Punjabi peasant who may struggle to any length to live and enjoy life. But the present situation seems to cross all limits of tolerance such that peasants have become hapless and see no way other than killing themselves. A pathetic situation indeed, for a people who are reputed to be brave, industrious and adventurous.

The problem of indebtedness has always been with the peasantry—big and small—though for different reasons. The difference is of the extent and severity of its consequences. In his attempt to look into the origins of debt, Nazir finds that the Punjab peasantry was never free from debt but the nature of assessment, definition of seminal concepts, mode of payment, and so on were crucial in defining the nature and extent of the problem. For instance, the problem of debt was never a serious one as it came out to be during the colonial period. Nazir argues that the land revenue under the Sikhs was normally collected in kind and rarely in currency, and the system of payment too was elastic and flexible. But with the advent of the British rule it was completely reversed. He writes:

> there is unanimity among scholars that the land revenue system in this era was not a cause of endemic rural indebtedness leading to alienation. This was because the land tax, such as it was, was normally not raised in money but in kind; and it was also proportionate to the crop production in good as well as in bad years . . . but it is important to stress that endemic rural indebtedness was virtually unknown under the Sikhs. (Nazir, 2000: 57–8)

What is important here to notice is that it is the political will of the powers that be that matters. Nazir notes: 'The various money-lending groups generically known as the *Sahukars* were *neatly kept in their place as servants* of the agriculturists' (ibid.: 60) (emphasis added).

After the annexation of Punjab in 1849, new rules and laws were implemented to favour the regime that led to high rise in indebtedness, mortgaging, alienation of land, its sale and purchase and litigation, and so on. Besides expansion of the area under cultivation due to canal irrigation and expansion of the global market in wheat and cotton, there was enhanced circulation of money. Nazir notes: 'The result for the peasants was catastrophic, and was recorded officially in every district' (ibid.: 65). The British changed the customary laws that vested land in the community or the village to individual proprietorship. This redefinition of the individual-land relation led to increased alienation of land which could now be sold to repay debts, something that had no scope under the former practice (of local laws). This way the British favoured the moneylender against the peasant-debtor that led to manifold increase in litigation. Thorburn records 96,222 suits disposed off for money in 1866; 1,44,956 in 1870; 1,77,279 in 1874; 1,91,635 in 1878 and 213,895 in 1883 respectively (ibid.).

They also introduced the new judicial system to facilitate the administration to implement the changed revenue laws. The increased circulation of money also enhanced the prices of land, thus making it a commodity worth investing in. Calvert shows that between 1866–74 the average annual sale of land was 88,000 acres, but it shot up to 1,60,000 acres during 1880–4 and further to 3,38,000 acres in 1890–4 respectively (ibid.). Thus, lending money to peasants, with land as surety, supported with the legal sanction of confiscating it in case of non-payment of debt prompted the non-agriculturist professional moneylenders to safely invest in this business. This investment was so lucrative that moneylending proliferated to such an extent that need arose to enact the Punjab Alienation of Land Act of 1901. It is interesting to note that these developments in the revenue system carried out in the second half of the eighteenth century had implications for modern Punjab on both sides of the border.

At the turn of the last century, Darling too confirms the poor plight of the peasant in South Asia: '. . . he is generally in debt, often hopelessly so. This is largely because he is improvident and lives from hand to mouth and rarely troubles to save' (Darling, 1977: xx). And, in India 'millions of cultivators are "born in debt, live in debt, and die in debt"' (ibid.: xx). The situation in this land of the five rivers was no different where more than 80 per cent peasants, both rich and small, were indebted. Darling writes: 'It is clear that, though the larger proprietor is more highly indebted, the smaller is more heavily involved' (ibid.: 12). But he never mentions a case of suicide by an indebted farmer. What has happened now that the psychologically secure, physically hardy and industrious Punjabi peasants are being pushed towards suicide?

The forces of the modern, ever expanding market seem to be the culprit. It was lacking a century ago. Lenin wrote that when a small farmer from a remote village brings his produce to the market, he gets linked with the international market. This linking has now become more stringent and foolproof. No one can escape it. Eric Wolf writes:

> . . . land and other resources were increasingly converted into commodities. . . . As commodities they were subjected to the demands of a market which bore only a very indirect relation to the needs of the rural populations subjected to it. Where, in the past, market behaviour had been largely subsidiary to the existential problems of subsistence, *now existence and its problems became subsidiary to the market*. . . . These various pressures also disrupted the precarious ecological balance of the peasant society. (Wolf, 1971: 266) (emphasis added)

The rise of the World Trade Organization (WTO) and its tightening noose is making all nations fall in line. What will happen to the small peasantry is anybody's guess. It has not done any good to the peasantry so far. Bhalla affirms: 'It appears that the introduction of economic reforms and becoming the founder member of WTO have not brought about the promised benefits to a large section of the poor farming community in India' (Bhalla, 2004: 310). Under such circumstances when the WTO insists on the withdrawal of subsidies, how can the small peasantry survive? Can we do away with the subsidies and yet compete with the farmers of the First World? Bhalla writes:

> But there is a clear message for the farmer. He has to come out of the mindset of dependence on ever increasing administered prices and on increasing subsidies. He will have to compete in the international market through increased productivity. (ibid.: 298)

This is easier said than done. It seems a Herculean task when we look at the ground reality. In Punjab at least there is hardly any cultivable land that is not under cultivation; irrigation has also reached the point of saturation; the water table is falling at a fast rate; and whatever intensive cultivation is possible under the existing conditions has already been exhausted. Only biotechnological research in agriculture can help enhance agricultural productivity in Punjab. But this opportunity has already been grabbed by the developed countries and their multinational corporations. How can we compete with them in research and development when funds for research and education in India are dwindling every year? Our premier Indian Institute of Science has a total budget of a mere Rs. 100 crore per annum while Stanford University annually spends Rs. 10,500 crore for research alone and Harvard University

Rs. 10,300 crore. Moreover our farmers are already under heavy debt. How can they look forward to buying and sowing genetically modified varieties of seeds? Will Monsanto and other MNCs forego their profits to help the poor peasantry make a living? Such questions loom large on the face of the Punjab peasantry, among many others.

The tightening noose of the capitalist market has already been experienced by two villages of Punjab—Harkishanpura in Bathinda and Shergarh in Muktsar districts. The panchayats of these villages have put their 'villages on sale'.[20] I wonder if the small peasantry anywhere in the world has taken such a distressful step, a collective decision by the whole village to sell not only the means of their livelihood but also the 'mother earth' with whom a peasant is spiritually attached. Which is a symbol of his prestige and honour. A 'village on sale' is a metaphor for the dying small peasantry.

It may be inferred from the above discussion that the problem of indebtedness is not new to the peasantry anywhere anytime, it is a normal fact in the Durkheimian sense of the term. But it is the prevalence of suicide amongst the peasantry that is a pathological fact and a recent phenomenon indeed. This problem has transcended the barriers of caste, region and religion. It is not only in Punjab that this issue has assumed greater seriousness but equally so in Gujarat, Maharashtra, Andhra Pradesh, Kerala and Karnataka (Shiva and Jalees, n.d.).

Besides the factors discussed above, the role of information technology, particularly television, has proved more suicidal for the peasant society. Though launched in the name of ushering an economically backward society into an era of modernization, I wonder if it has done any good to the rural Punjabi society except inducing the innocent folk to the glamour and the vices of urban society. More than being an instrument of modernization it has played in the hands of the market and media giants spreading crass consumerism. This has become more conspicuous with the liberalization of the market and free flow of information through satellites. Now, even the poorest of the poor peasant must have an 'idiot box', as a symbol of his social status and of being linked to the modern world. But, as every dark cloud has a silver lining, along with other forces of modernization, it has also helped the weaker sections of society become conscious of their rights. The women and the Dalits in particular have realized their potential to change the society and are striving towards their empowerment, which has made the latter especially confront their Jutt masters head on. The Jutt–Dalit tensions and confrontations in rural society are not only gradually becoming more widespread but also growing more fierce and violent all over the state.

This is a synoptic view of the problems that are afflicting the Punjabi peasantry, especially on the eastern side of the border. Its western counterpart too seems no better, barring aside the rich peasantry that constitutes a small proportion of the total population. The scholars from west Punjab shall deliberate on their problems at length. It would provide an interesting insight into the comparative problems of the peasantry on either sides of the international border operating under different systems of political economy and religious persuasions but largely similar socio-cultural conditions.

In west Punjab, the size of the landholdings are extremely large but concentrated in a few hands. It was the need of the imperialists to concentrate land and authority in a few hands. Saif notes: 'Establishing a landowning class in Punjab is significant because of "colonial recruitment policy of Indian army". In the aftermath of the great revolt of 1857, Punjab emerged as a major recruiting area of the colonial army' (Saif, 2003: 101). Talbot also asserts:

> The British interest in harnessing the waters of Punjab and Sindh notwithstanding, their advance in Muslim north-west India was prompted by military rather than commercial considerations. . . . Punjab was thus seen as a base for military operations. (ibid.: 102)

And, to safeguard their colonial interests, says Sayeed, the British introduced a system 'which generated the worst kind of feudal exploitation of the Muslim peasants' (ibid.: 101).

At the other extreme is the small peasantry whose socio-economic conditions have not improved significantly despite the 1970s' Green Revolution. Their plight is worse than the Punjabi peasant on the eastern side of the border. The main reason for these skewed agrarian class relations is the presence of strong feudal lineages and heritage and non-implementation of modern land reforms. The absence of political democracy and low literacy level of the general population have made things worse for the peasantry.

The land reforms were first implemented in Pakistan in 1959 but without the political will to implement them in letter and spirit. Consequently the ruling elite including the military and civil bureaucracy tilted scales of the reforms in their favour.[21] Akbar Zaidi comments on the 1959 reforms: 'what they (land reforms) did was to distribute power away from some landlords and include the civil and military elite in their strategy' (Arif, 2004: 21). Consequently, the military officers became a new breed of absentee landlords. Not only that, Hussain notes: 'In Pakistan, the military runs the largest real estate operation in the country' (ibid.: 22). Unfortunately,

> Land reforms, upon implementation, actually increased rural poverty as big landowners turned to self-cultivation through hired farm labour and mechanisation, and ejected their traditional tenants. These tenants lost their livelihood, shelter, material and social assets, often ending up as unskilled destitute in big cities. (ibid.: 1)

The collusion between civil and military bureaucracy has become the cause of peasant uprising at the turn of the century. The Punjab government owns more than 68,000 acres of land cultivated by tenants since the early 1900s. Out of this, 26,274 acres are with the military and 7,185 acres with the Punjab Seed Corporation. In both the cases large parts of the land are cultivated by tenants. 'These agencies' plan to change the tenure arrangement (tenancy) to a contract system sparked a tenant uprising that has now become a resistance movement in Pakistan' (Arif, n.d.: 3). It is spearheaded by a million-strong peasant organization called Anjuman Mazarain Punjab.

Khan et al. examined the last two agricultural censuses of Pakistan and found that (i) large farms (60 acres plus) accounted for a negligible number of total farms in both 1980 and 1990, but the area cultivated by them has more than doubled. (ii) the middle farm size has thinned out, and there is greater cultivation by the smallest and largest farm-size categories (Khan et al., 2001). In 1990, farms less than or equal to 12.5 acres (according to Siddiqui, up to 12.5 acres of irrigated land is considered below subsistence level) accounted for 80.6 per cent of the total cultivated area, while this has increased to 88.2 per cent in 1998–9.[22]

The economic plight of this class is as bad as its counterpart in east Punjab. These peasants are poor and survive only on debt that is rising annually.

> In 2000–1 small farmers owning land up to 12.5 acres (5.06 hectares) borrowed more than 15 billion rupees from the Agricultural Development Bank of Pakistan, whereas owners of large holdings (above 100 acres) got only 443.2 million rupees. . . . (Arif, 2004: 18)

Before the drought years in 1996–7 the total loan to small farmers amounted to Rs. 4 billion, which doubled to Rs. 8.5 billion in 1997–8 and jumped to more than Rs. 14 billion in 1998–9 (ibid.).

> Indebtedness of all households to non-institutional sources of credit increased from Rs. 27.9 billion to Rs. 54.6 billion while that of the farm-households more than doubled from Rs. 17.5 billion to Rs. 37.5 billion, raising the debt ratio of total indebtedness from 61 to 68 per cent. (ibid.: 19)

A reader may question the greater emphasis on the history, religion and culture of the east Punjab peasantry than on its western counterpart. It is so for the simple reason that a larger number of papers focus on the problems on the eastern side. Since the roots of the present are in the past, many intricate problems of the present can only be comprehended if one situates them in its socio-economic, historical and cultural contexts. The whole idea of providing comprehensive details of the historical landscape is to make sense of the present in terms of its past leanings. No doubt the ambition of the organizers was to include all contemporary problems afflicting the peasantry but for various reasons this ideal has remained unrealized. But still no significant problem has been ignored be it the case of globalization influencing the Punjab peasantry or the national policies of our own government. The problem of suicides is most crucial and pathetic as is the 'forced' migration of youth. The negative consequences of the Green Revolution have also been discussed as is the problem of militancy/violence in the recent past. The farmers are now representing themselves through various unions and organizations, yet there is no thaw in their socio-economic problems. Rather the farmers' movements have themselves been afflicted with the viruses of factionalism and populism, thus derailing them from their primary objectives. Such are the issues that have been discussed by various experts in their own areas of research with the sole design of projecting the plight and problems of the peasantry such that some solutions could be gleaned out of the dark tunnel.

The multifaceted problems of the Punjab peasantry cannot be addressed in one conference but we have tried to deliberate upon as many aspects as possible. Another important feature of this conference had been that the peasantry's turmoil is not confined to one or two disciplines or approaches but numerous ones. The delegates and the authors to this volume include the litterateurs, sociologists, economists, political scientists, activists in the farmers' movement, theatre and TV artists and other freelancers.

Since we have also attempted to look into the situation of west Punjab (Pakistan) it would not be out of place to begin with the profile of a village there as given by Ahmad Salim. It is an exercise in ethnography, giving details of class/caste relations that have only become worse with the introduction of a liberal economy. The history of the family and the village community is vividly described, explaining the nature of class and class conflict historically in the style of a litterateur. The lower castes and classes are now being subjected to new forms of exploitation. The situation in east Punjab villages is of course more relaxed, thanks to the teachings and practices of the Sikh religion and also democratization of Indian politics.

The dimension of history, so very crucial in contextualizing the present, is provided by Lubna Saif. She traces the impact of colonial capitalism on Punjab agriculture. Her focus is on west Punjab, but that was the 'real' Punjab before the Partition of this subcontinent. Thus this paper elucidates the historical underpinnings of the phenomenon that helps better comprehension of the present-day problems of the two Punjabs.

One of the most important factors responsible for tightening the noose on the farmers, pushing them to commit suicide, is the process of globalization which is seen as nothing different from capitalism in its neo-imperialistic mode. Aijaz Ahmad situates the peasantry's turmoil in the crisis of global capitalism. He well documents the process by which globalization has gradually gained acceptability in the developing countries the world over.

Utsa Patnaik tries to explain the crisis in peasantry owing to the adoption of the new liberal policy by the Government of India. She says it in very strong terms by referring to it as the devastation of India's granary. She convincingly argues how the neo-liberal policy has systematically destroyed Punjab's agriculture.

For Staffan Lindberg, the issue of farmers' collective action is getting more complex with the onset of globalization. Who is the enemy and how to fight with him? Is it possible to deal with the 'global' enemy in the classical mode? Lindberg also suggests a way out for the farmers. He advises them to launch a collective action in the form of cooperatives that should market products too.

Similar advice is given by Sukhpal Singh. He argues that until and unless farmers adapt 'new generation cooperatives' that are market oriented there is no respite for the farmers. The increasing costs of agricultural technology can be partially mitigated by encouraging contract farming by private corporate business interests. The State and the NGOs can also intervene to protect the interests of the farmers and the broader community. Singh specifically analyses the role of the institutions and organizations in tackling or perpetuating the crisis.

Zia ur Rehman from Pakistan advises against corporate farming as it is being implemented in west Punjab. He fears that poor peasantry will not be able to match the capital might of the multinational corporations. The Government of Pakistan is out to help the MNCs in agribusiness that would further push the poor farmer to absolute poverty. On the contrary, Rehman argues in favour of traditional and sustainable agriculture as a viable alternative. This would ensure more food security to the subsistent farmers.

Balkar Singh and Pashaura Singh hold the Green Revolution responsible for the pitiable condition of the peasantry. They provide an activists' account of the crisis without academic jargons as both of them are actively engaged in the peasant movement.

The fate and state of contemporary peasant movements referred to by Lindberg is discussed in detail by Sucha Singh Gill. He attempts to explain the crisis of viability of the small and marginal farmers that has deepened over the years and the indifference of the farmers' movement in resolving their problems.

The dimension of violence that engulfed east Punjab during the 1980s is also a manifestation of the deepening economic crisis according to Birinder Pal Singh. He argues that the Sikh militants were not using violence as an end in itself but as a means to an end—for registering their protest in its extreme form as all other means, according to them, had failed.

If the Sikh militants adopted violence as a form of protest and as a means of claiming their rights, the Muslim peasants in west Punjab chose the path of non-violent protest to stop their eviction from their tilling plots by the Pakistan Army and the government. Aasim Sajjad Akhtar provides a detailed account of the Anjuman Mazarain Punjab's struggle over the last decade.

To Ronki Ram, the growing Jutt–Dalit conflict is also a manifestation of the agrarian crisis. He argues that Dalits who have historically been subjected to domination and exploitation are now asserting themselves in the wake of growing democratization and consciousness of their civil rights. Their education and political mobilization is witnessed in the growing clashes between the two communities that form an integral part of rural Punjab.

Darshan Singh Tatla looks at the problem of increased migration of Punjabi youth to the developed countries, even at a great risk to their lives, as a symptom of the peasantry's crisis. He argues that it is not only the failure of economy that is pushing the youth to greener pastures but also the failed state. The lack of political will on the part of leaders, increased corruption and weak administration have created a situation of anomie. Thus, the youth are forced to migrate to ameliorate their living conditions.

As mentioned above various solutions to this crisis have been suggested by different authors. One very important measure that is being trumpeted as the sole panacea for the peasant crisis is breaking down the wheat-paddy cycle and venturing into crop diversification. The Punjab government even appointed the Johl

Commission to this effect. But this report is replete with flaws which is why the farmers are not going in for crop diversification. Harjinder Singh Shergill documents the limitations of this scheme in his critique of the above report. He argues against the suggested policy of diversification and informs that under the given conditions wheat–paddy cropping is the best bet for the farmer.

Jodhka provides an overview of the agrarian crisis in the state cautioning that talking about the existing state of affairs in terms of 'crisis' could be elusive and could distract us from understanding the social and economic changes as on the ground. Thus, he looks at such changes in rural Punjab and tries to conceptualize them in terms of the processes of liberalization and globalization. He argues that the village should be seen in relation to its surrounding environment as that would enable us to raise more meaningful questions about what is happening to agriculture now.

NOTES

1. It is often said that a Jutt/Jat first acts and then thinks. Spontaneity and primacy of action at the spur of the moment are his traits rather than a rational calculation of its consequences. Another oft quoted saying about them is: *Jutt ganna na devey, bheli de de* that a Jutt may not oblige someone with a stalk of sugar cane but may give him a lump of jaggery.
2. The geographical location of Punjab in the north-west India is the reason for the extreme weather conditions of this region. It has very hot summers, heavy monsoon rains, hot and humid months after the rains (*Bhadon di garmi da marya jutt sadh ho jandai,* that Bhadon's [September–October] hot and humid weather forces a peasant to renounce the world and become a mendicant), then autumn leads to harsh winters followed by a short span of spring (Basant).
3. Three acts that accelerated the pace of land reforms in the state are: the Punjab Security of Land Tenures Act 1953, the PEPSU Tenancy and Agricultural Lands Act 1955 and the Punjab Land Reforms Act 1972.
4. No doubt the Jutts dominated but the lower castes also threw up their chiefs who were no less important. In the selection of its chief or *Sardar,* preference was given to 'suitability over hereditary claims and caste distinctions'. Bhagat Singh (1978), *Sikh Polity: In the Eighteenth and Nineteenth Centuries,* Delhi:

Oriental Publishers, p. 94. For instance, a Jutt could also lead a Ramgarhia *misl.* All the *misls* combined to form the Khalsa or Sikh commonwealth. Also see J.D. Cunningham (1972), *A History of the Sikhs*, Delhi: S. Chand and Co., pp. 96–9.

5. The era prior to Ranjit Singh was charged with the Khalsa ideology to such an extent that Sinha writes: 'He did not assume the title of King but rather the impersonal designation of *Sarkar* to denote the source of orders' (quoted in Banga, op. cit., p. 37, fn. 105). Amar Nath also tells that 'Ranjit Singh did not approve of the epithet *hazur* for himself' (ibid., p. 37, fn. 107). But Banga quotes the Pindori Documents to show that 'the term *khalsaji*, whether used with *sarkar* or *darbar*, did not refer to the Sikh people. It referred only to the individual ruler in question, whatever his title' (ibid., p. 37).

   An epithet of Maharaja is used here because this is how Ranjit Singh is known in history books and others. He is almost always addressed as *Sher-e-Punjab* (Lion of the Punjab) Maharaja Ranjit Singh whose period in Sikh history is considered the golden age in popular chronicles and books.
6. In Punjabi this also means treasure which of course, cattle are to a peasant.
7. The area under rice cultivation increased from 11,83,000 hectares in 1980–1 to 20,15,000 hectares in 1990–1. The corresponding figures for wheat are 28,12,000 and 32,73,000 hectares respectively. The production of rice was 32,33,000 metric tonnes in 1980–1 and 65,06,000 metric tonnes in 1990–1 and the corresponding figures for wheat were 76,77,000 and 1,21,59,000 tonnes respectively. *Statistical Abstract of Punjab 2004*, p. 164.
8. The total cropped area shows a gradual decadal ascent from 1960–1 to 2000–1 as 4,732, 5,678, 6,763, 7,502 and 7,941 thousand hectares respectively. *Statistical Abstract of Punjab 2004*, p. 159.
9. There were 1.92 lakh tube wells in 1970–1; 6 lakh in 1980–1; 8 lakh in 1990–1; 10.73 lakh in 2000–1 and 11.68 lakh in 2004–5. *Statistical Abstract of Punjab 2005.* Official figures are not available since then.
10. There were 1,18,845 tractors in 1980–1; 2,89,064 in 1990–1; 4,34,032 in 2000–1 and 4,59,424 in 2003–4. *Statistical Abstract of Punjab 2004*, p. 587.
11. Gross irrigated area also recorded a rise over the decades 1960–1 to 2000–1 as 4,242.5, 5,781.3, 7,054.8 and 7,663.8 thousand hectares respectively. *Statistical Abstract of Punjab 2004*, p. 240.
12. The number of schools (from primary to class 12) has grown consistently over the last four decades. In 1971 there were 9,641 schools that grew to

15,950 in 1980, 16,580 in 1990, 18,998 in 2000 and 19,682 in 2003. In 1971 there were three universities, two engineering, four medical and 122 degree colleges in the state. These numbers have shot up to 5, 16, 6 and 209 colleges respectively in 2003. The trend has not slackened since then, rather has became more pronounced with increasing privatization of education in the era of liberalization. There is a glut of establishing such types of professional and non-professional institutions in the state and more so in the rural areas to avail subsidies. These projects ensure good profits to the investors. There are numerous such cases where merchants and commission agents with no expertise in education are running these institutions. The teachers are paid the least and the students are charged exorbitant fees. Currently there is no business more profitable than this.

13. In 1980 there were 4,25,046 unemployed persons registered in Punjab of which half were educated (matric and beyond) and half uneducated (below matric and illiterate). The total number in the subsequent years were 5,37,330 (1985); 6,59,250 (1990); 5,06,236 (1995); 5,36,159 (2000); 4,63,229 (2005) (*source:* Director Employment, Punjab Government, Chandigarh). The decline from the highest 5,36,675 in 2001 touching 4,54,335 in 2006 does not indicate that more persons have got employment, but that people have stopped registering themselves in the Employment Bureaus as there seem no chances of their getting employed even if vacancies are advertised. The government's policy of freezing sanctioned posts, high corruption (both political and economic) and loss of faith in the recruitment process are some of the salient factors. The actual number of unemployed persons is estimated to be more than double.
14. The infamous Malta boat tragedy refers to an accident on 25 December 1996 in which an overloaded boat carrying illegal immigrants capsized in the Mediterranean Sea drowning over 290 persons mostly from India, Pakistan and Sri Lanka. Of the 170 Indians who died, most were from Punjab. The parents and relatives of the victims have formed a Malta Boat Tragedy Probe Committee to claim compensation.
15. Migration to the developed countries by the poorer strata is due to push-factor but in the upper classes it is due to the pull-factor. These classes do not foresee a bright and secure future for their children due to high unemployment, high fees especially in the technical institutions and an overall corrupt socio-political system in which an honest person feels helpless and rudderless. The present-day middle classes are more given to this trend. The well established intelligentsia are leaving their jobs for the children's future.
16. A former Member of Parliament and president of the Lok Bhalai Party,

Balwant Singh Ramoowalia has records of more than 10,000 such cases. He along with others has taken up this matter with the Governments of Canada and the United Kingdom to take such measures that may restrain the proliferation of such cases. There are reported cases where close relations including real brother and sister have been married on paper to facilitate immigration.

17. *Statistical Abstract of Punjab 2004*, p. 156, Chart No. 3. 'Land Utilisation in Punjab 2003–4 (P)'. Elsewhere in the *Statistical Abstract of Punjab 2004* the said figure stands at 6.12 per cent (p. 285). The *Abstract* clarifies that the gap in these figures is due to the fact 'that certain lands though not wooded, are taken as Forest by the Forest Department while these are not treated as such by the Director of Land Records', Punjab. *Statistical Abstract of Punjab 2004*, p. 155.
18. The consumption of chemical fertilizers—nitrogenous, phosphoric and potassic, that is, NPK—has taken leaps from 2,13,000 nutrient tonnes in 1970–1 to 7,62,000 in 1980–1; 12,20,000 in 1990–1; 13,13,000 in 2000–1 and 15,43,000 in 2003–4. *Statistical Abstract of Punjab 2004*, p. 215.
19. The case of *Amrikan sundi* (American bollworm) on cotton in the Malwa region is well known. Even eight to nine sprays of powerful pesticides did not kill the pest. The high use of pesticide in the cotton belt over the years has led to high incidence of cancer amongst the inhabitants of this region. The Post Graduate Institute of Medical Education and Research, Chandigarh, and the Centre for Science and Environment, Delhi, have separately brought these facts to light.
20. Harkishanpura is a village in the Rampura Phul block of district Bathinda with 141 households. There are only 21 families with only 8 acres of land that do not have any loan standing in their name. All others have debts varying from less than Rs. 0.25 lakh to more than Rs. 10 lakh. The latter has only four families. The greatest concentration is in the range of Rs. 2–5 lakh. There are 35 families owning 251 acres of land with the loan amount of Rs. 126.14 lakh standing in their name. *Survey Report up to 25.10.2002 of village Harkishanpura, Block Rampura (Distt. Bathinda)*, Bathinda: Nodal Agency, D.R.D.A., p. 6. Also see '*Pind Vikau Hai*' (in Punjabi) *Desh Sewak, Aitvari Magazine*, Chandigarh, 8 September 2002.
21. Three land reforms were carried out in Pakistan in 1959, 1972 and 1977. Each time the land ceiling was fixed on individual basis rather than the household. In 1959 it was fixed at 150 acres of irrigated land, 300 acres of semi-irrigated and 450 acres of un-irrigated land. In 1972 the ceiling on irrigated land remained the same but that of un-irrigated land was lowered

to 300 acres. In the 1977 reforms irrigated land ceiling was lowered to 100 acres and un-irrigated to 200 acres. These limits clearly reflect the bias against the small peasantry.

22. Ibid., pp. 2–3. According to Kamal Siddiqui, subsistence level begins at 12.5 acres of irrigated landholding and ranges up to 25 acres; 25–50 acres are above subsistence but below economic (i.e. profitable) level. The economic level ranges between 50–64 acres while 64 acres plus accounts for above economic level (Arif 2004, p. 18).

REFERENCES

Arif, Mazhar (2004), *Land, Peasants, and Poverty: Equitable Land Reforms in Pakistan*, Islamabad: The Network Publications.

——— (n.d.), *Peasants in Revolt*, Islamabad: The Network for Consumer Protection.

Banga, Indu (1978), *Agrarian System of the Sikhs: Late Eighteenth and Early Nineteenth Century*, Delhi: Manohar.

Bhalla, G.S. (2004), *State of the Indian Farmer: A Millennium Study*, vol. 19, *Globalization and Indian Agriculture*, Delhi: Academic Foundation.

Darling, Malcolm (1977), *The Punjab Peasant in Prosperity and Debt*, Delhi: Manohar.

Habib, Irfan (1976), 'Jatts of the Punjab and Sind', in Harbans Singh and N. Gerald Barrier (eds.), *Punjab: Past and Present: Essays in Honour of Dr. Ganda Singh*, Patiala: Punjabi University.

Khan, Shahrukh Rafi et al. (2001), *The Case for Land and Agrarian Reforms in Pakistan*, Islamabad: Sustainable Development Policy Institute.

Nazir, Pervaiz (2000), 'Origins of Debt, Mortgage and Alienation of Land in Early Modern Punjab', *Journal of Peasant Studies*, vol. 27, no. 3, April, pp. 57–8.

Pettigrew, Joyce (1975), *The Robber Noblemen: A Study of the Political System of the Sikh Jats*, London: Routledge and Kegan Paul.

Ray, Niharranjan (1975), *The Sikh Gurus and the Sikh Society*, Delhi: Munshiram Manoharlal.

Saif, Lubna (2003), 'Impact of Colonialism on Land Holding and Authority—Causes of Underdevelopment in Pakistani Society', *Pakistan Perspectives*, vol. 8, nos. 1 and 2, January–December, pp. 83–110.

Shanin, Teodor (1971), 'Peasantry as a Political Factor', in T. Shanin (ed.), *Peasants and Peasant Societies: Selected Readings*, Harmondsworth: Penguin.

Shergill, H.S. (1998), *Rural Credit and Indebtedness in Punjab*, Chandigarh: Institute for Development and Communication.

Shiva, Vandana and Kunwar Jalees (n.d.), *Farmers Suicides in India*, Delhi: Research Foundation for Science, Technology and Ecology.

Singh, Birinder Pal (2005), 'Some Comments on the Future of Rural Development in India', *Gandhi Marg*, July–September, vol. 7, no. 3, pp. 187–200.

Singh, Ganda (1956), *A Brief Account of the Sikhs*, Amritsar: Shiromani Gurdwara Parbandhak Committee.

Statistical Abstract of Punjab 2004, Chandigarh: Government of Punjab.

Thorner, Daniel (1971), 'Peasant Economy as a Category in Economic History', in T. Shanin (ed.), *Peasants and Peasant Societies: Selected Readings*, Harmandsworth: Penguin.

Wolf, Eric (1971), 'On Peasant Rebellions', in T. Shanin (ed.), *Peasants and Peasant Societies: Selected Readings*, Harmandsworth: Penguin.

CHAPTER 1

Class Conflict and Change: Profile of a Village in West Punjab

Ahmad Salim

While speeding on the grand motorway from Islamabad to Lahore, you may pull out for a while at Miana Gondal just a couple of miles away from Salam Interchange. It is a small village, doing its best to catch up with the mechanized world. Earlier, it was a part of tahsil Phalia in the district of Gujrat. But now it is a part of tahsil Malakwal of district Mandi Bahauddin.

NEW SCENARIO/PERSPECTIVE

Miana Gondal is an old village in the interior of Punjab. However, it can no longer be termed as a typical village in the classical Marxist sense because it has lost its self-sufficiency and has gotten close to the city by harnessing the latest marvels of science. Though the story of its urbanization stretches over a mere two decades, it has shaken Gondal's stillness and calmness, which had always been a typical feature of a classical village. In the early part of the nineteenth century, a dispensary, a post office and a primary school for boys were established. After the inception of Pakistan, the boys' school was upgraded to the middle level and a primary school for girls was also started. By now, the girls and boys both pass their matriculation there. Recently, an intermediate college for boys has also been approved by the government that is yet to be established. Perhaps, the influential residents of the nearby village, Bosal are trying hard to establish the college there. However, all these facilities are specific for the children of landlords, worthy shop owners and the well-off Khwaja families. The educational opportunities for the lower strata of the village, especially the Muslim sheikhs known as *musallis* are very poor.

Geographically, Miana Gondal is situated in the suburbs of Mandi Bahauddin and is adjacent to Phularwan and Bhalwal, the two towns in Sargodha district. The other side is linked with tahsil Pind Dadan Khan of district Jhelum. The village is 73 miles from Gujrat, 40 miles from Sargodha, and 143 miles from Lahore, the capital of Punjab. Despite its distance from the major cities, the village has developed a civic air of its own. After the completion of the motorway, it is about two-and-a-half hour drive from either Lahore or Islamabad. All the modern technology and equipment considered to be the hallmark of a city are seen in Miana Gondal. One will find transistor radios, tape recorders, televisions, dish antennae, telephones and all the latest sophisticated agricultural tools and equipment. However, these things have only given the residents a tinge of materialism, shallow social status and frustration.

Things have changed now. Television programmes have replaced the tradition of puppet shows in the landlord's house. People are the same but their occupations have changed. The new generation of weavers is preparing sweets or selling vegetables. The specialist artisan groups too have changed their occupation due to modernization.

The boys and girls with whom I spent my childhood playing games around the village now laugh at my 'conservative' approach and feelings. I might be too romantic about the past, but I felt sad to see all the good traditions of my village being dashed to the ground. Now, there is only a blind race for money. Money, mosquitoes and houseflies are increasing just as the filthy ponds and heaps of garbage mount day-by-day.

Not too long ago, it was only a small section of the upper class of the landlords or Khwaja family who exploited the low class villager. Now the rising class of shopkeepers has joined hands with the landlords in their exploitation. In the last couple of years, many new and modern houses have been constructed. These houses are especially comfortable because they have flush toilets and a regular supply of water. However, nobody has ever thought of changing the negative attitude toward the lower classes.

No doubt, new inventions have upgraded the standard of living in Miana Gondal but only to the consternation of the lower class. Their fate is still the same even though they are changing their occupations in an effort to join the ongoing material race. The traditional baker, Old-ma Phattein's (Fatima's) common mud oven (*tandoor*) is no longer in a working condition and people do not gather there to get their breads baked. Old-ma's third generation is sweating blood to have a decent life than ever before. Uncle Yousaf, Old-ma's older son, is exerting more than before to keep his body and soul together. On the other hand, his younger brother, Nazir, who had always been proud of his dandy horse and tonga, now looks for new openings such as selling ice to earn a living.

In one corner of the village lives a group of the degraded outcasts called *musallis* (Muslim sweepers) in disparaging conditions. These *musallis* used to clean the households and carry garbage from the houses of the Khwajas and other families for a petty wage of Rs. 10 per month. After 1980 their plight has worsened. The brick kiln owners exploit the *musallis.* Nobody sympathizes with them or does something concrete to ameliorate their inhuman condition. Instead, they are verbally insulted and degraded by the members of the upper classes.

At the beginning of the twentieth century—half a century before my birth—the Khwajas introduced materialism to this village.

KHWAJAS: THE EARLIEST SHOPKEEPERS

The first house my grandfather Mian Fazal Karim bought in Miana Gondal is situated in front of the small graveyard and the village market. On one side of the house live *miyanas* (religious/spiritual leaders) and on its back reside the blacksmiths whose oldest member, uncle Ghulam, is a living history of the village. Another senior member, Muhammad Yousaf, of the same clan, recently died of cancer. He was not provided adequate treatment at the Shaukat Khanum Hospital for lack of funds. However, Muhammad Yasin, Shafiq, and their children (third generation) still talk earnestly about their past links with our family but the warmth is not reciprocated by the members of our family. Now, nearly all of my grandfather's family has shifted from the village to Lahore, Islamabad and Mandi Bahauddin. They have broken their ties with the village. During my childhood, it was puzzling how men in our family could go to the blacksmiths' houses and talk to their women, but their male members were neither allowed to enter our houses nor talk to our women. They came to our house only when there were no women in sight.

It was from the same uncle Ghulam that I learned a quarter of a century ago that my grandfather migrated from Pind Dadan Khan in district Jhelum to this village. In those days, Muslims were setting up small shops in competition with the Hindu shopkeepers. My grandfather was also among the pioneers of Muslim trade. Being a blacksmith, uncle Ghulam is the appropriate narrator for this transitional period since it is directly concerned with the trade sources and tactics that replaced barter system and introduced money as a medium of exchange. This also introduced the concept of wage labour for cash in the *musallis*, initially in our houses, but later on by big landlords and contractors of brick-kilns. Uncle Ghulam reminisced:

> Your grandfather, Mian Fazal Karim, came to Miana Gondal and said, 'I have to set up a shop here.' Linen and muslin were not in vogue here then. There used to be cloth, woven in looms. There was no system of yards or metres. Instead cloth was measured by hands; four to six hands of cloth were enough for a skirt.
>
> The same cloth was usually coloured black for turbans. Mian Fazal Karim opened a shop here and introduced linen and muslin. He opened his shop first in the place where the goldsmith Sattar sits. The narrow-width of the loom-woven cloth meant nobody knew how much of it would be sufficient for a shirt or a turban. Hayyat Dhabba's (tailor) father, Muhammad used to stitch clothes at that time. People normally went to consult him regarding the exact length of cloth required for their clothes. However, now it was a source of constant confusion to them. Your grandfather, one day, went to Miani (district Sargodha) and fetched a tailor, Imam Din for his shop. This lowered the importance of Muhammad.

Uncle Ghulam also shared a long story about the differences between the local and immigrant tailors. The local tailors stitched clothes in exchange for goods whereas the immigrant tailor charged money for the same work. The latter also introduced modern fashion in stitching. So my grandfather's cloth shop hit the looms hard and his modern tailor successfully introduced money, trade and novel designs. The small shop also uprooted the old forms of production. The modern culture introduced by my grandfather flourished rapidly. But just as old colonialism was pushed back by American imperialism, this new trend and development also slipped away from the clutches of our second generation.

First, agricultural land saw its doom and then divisions within the household turned the shops into insignificant small units. This socio-economic dislocation occurred during my childhood. I still remember men and women coming to our house for washing our clothes, sweeping, carrying garbage away, cutting wood and fetching water. They were given our old and used clothes in lieu of payment. Some of them were also hired. They were asked to do a task but were made to do a number of other chores too, thus exploiting them. Their nominal wages were fixed but the sphere or nature of the job was unlimited. For example, Bahishtan Musallan was paid just Rs. 5 a month for sweeping and taking out the garbage of our house and of other Khwaja houses too.

MUSALLIS' SLAVERY: THE FIRST PHASE

The seeds of the exploitative system sown by my grandfather slowly started blossoming. Currency was in common use. First, the silver coins and later the paper currency (rupees) started showing their magic. This had great charm for the *musallis.* I remember that I used to refer to Bahishtan, the *musalli* woman who swept our home, as a *Phuphi* (paternal aunt, that is, father's sister). We belonged to the upper-middle class and had emigrated from a bigger town. Our manners and courtesies were characteristic of the 'civilized' people. We were 'cultured'. We knew that we could get them to carry our garbage for just Rs. 2. Ironically, this was an act of large-heartedness on the part of the high caste people. We felt we were keeping the worthless low-class people under our patronizing care.

In those days, there were only 10 or 12 Khwaja families in Miana Gondal. So the net wages of Bahishtan was never more than

Rs. 50–60 a month. At times, she could make more money if the guests in the house were generous to her. Whenever my paternal aunts visited us, they gave the servants some money as an appreciation of their services and faithfulness. It was also a symbol of status for our family, that our guests too showered such patronage on the menials. It showed: *Assin rajje-pujje han te sadde rishtedar vi* (that we are well-off and so are our relatives).

I was different from the rest of my family. After completing my primary schooling, I left my native place and went a couple of hundred miles away for further education, the first rebellion of its kind in the family. I stayed with a close relative and experienced the hardships of practical life. I saw the atrocities of imperialistic capitalism on the people of lower ranks of my village as well as elsewhere. I could empathize with the fate of the *musallis.* I thought about the chronic cruelties enforced on these people. I remembered the innumerable times my friends and cousins used the female servants to satisfy their sexual urge on the roofs, porches, basements and even lavatories of our house. It was a well-established belief of the town's 'gentlemen' that these 'cheap' servants were meant to please and serve them in every possible way.

I recall getting into a squabble with my elder brothers on this issue when I visited my village during vacations in 1985. I cried out: 'A *musalli* also has self-respect. S/he is a human being too, like us. They are also created by God.' My discourse on modernity did not enlighten them. Rather I made a fool of myself before them. They were ridiculing my arguments and laughing at my 'wisdom'. They also teased me, calling me 'a lad from the city'.

On another occasion while I was giving a lecture about *musallis,* I even scolded my brother 'not to talk nonsense about them'. This was considered an act of blasphemy on my part. The elders of my family and community chided me on this count. Ironically, a *musalli* youth got up in the course of the lecture and criticized me for being rude to my elder brother. It shows the extent to which slavery had permeated to the core of the *musalli* 'slaves'.

Our elders did not disapprove of the youngsters' sexual misconduct with *musalli* women. On the contrary they themselves didn't miss a chance to wink at them. This was considered their privilege, the right of the superior castes and classes. The women from lower castes and classes were considered playthings for their sons. An act of manhood hence applauded, though covertly. It was considered a problem and a 'moral lapse' on someone's part only when the indulgence exceeded limits. Then, the solution to such 'moral lapses' was to get the son married. But marriages too have failed in stopping this practice. Many youngsters have continued exercising their 'special rights' with respect to *musalli* women. The worst part of this story is that the wrongdoer was never accused of any immoral act; it was the victim herself who was blamed for her 'misconduct'.

BAHISHTAN: HER LIFE

I met Bahishtan in 1982 when she shared her life-story with me. Her earliest recollection of her family history was that her ancestors immigrated from Thakur Waryam to Miana Gondal. Here, they worked hard and being thrifty saved money too. They purchased a plot and erected a small mud-house where she lived until her death. She was about 80 years old when she died. With her death in the late 1980s her family moved to different parts of the country. Her youngest son Mutalli is now living in a single-room house. Bahishtan's fourth generation has also grown up. Her house, like those of other *musallis*', was situated on the farthest end of the village as she belonged to a menial caste. The distance was meant to mark the caste difference. She could not remember the exact price of the plot, but mentioned that it cost her Rs. 200 to erect two separate single-room houses, one of them was made of bricks from the kiln.

Bahishtan devoted all her life to the service of the Khwajas. However, with changing times, the Khwajas either built new houses or left the village. Bahishtan became jobless. Her sons began looking for work. Her eldest son, Sardara, started working at a brick-kiln, marking the beginning of a new kind of slavery in which a man puts himself on sale along with his wife and children to the owner or the contractor. Her second son, Karmalli also joined a kiln. The contractors put the labourers to tedious tasks. Bahishtan said:

First, they lend some money, which is repaid in the form of monthly deductions from the salary. As a result, we sell ourselves for years to come. The irony of the situation is that women and children also become their possessions. As long as the loan is not paid back, we are their slaves round the clock. It is true for all the kilns. Just look at the cruelty! They pay only Rs. 20–2 at the most for laying a thousand bricks. Half of the amount is deducted out of the total wages as installment of the loan and the other half is given to us to meet our daily needs. If this was all, we could have accepted it with some satisfaction. But they cheat us in accounts too. If they deal fairly, the whole amount of the loan could be repaid within a few months but the contractor keeps increasing the advance amount in a technical way such that we remain their slaves forever.

The brick kiln business is at its zenith in Haripur, Azad Kashmir, Jhelum and Rawalpindi. The contractors extort money, which is why every contractor has three to four kilns. They pay us Rs. 1000 but note it as Rs. 2000. So we fail to purchase our freedom even after the end of a season. We remain in debt more than before even after paying them back more than the due amount in the form of deductions from our wages. As a result, we are bound to stay their slaves even after the season is over and there is no work for us. We cannot go anywhere to earn our bread. Prisoners are never choosers. We remain prisoners along with our children, sisters and brothers. At times, men go to other kilns for labour but their women and children stay back as prisoners in the previous kiln. This is the most terrible state of affairs.

Take the case of Sardara who has gone for work at another kiln, leaving his wife and children behind. He will get some amount as advance from the contractor of the new kiln to liberate his wife and kids from the earlier kiln and become a slave of his second master. It is a heinous cycle of slavery that we cannot liberate ourselves from.

The *musalli* women of Miana Gondal have nearly given up household drudgery. They accompany their men to the brick kilns, as this is one of the conditions of the contractors. They insist on the company of women, obviously to prepare food for their men. The women cook and work along with the men in making dies for bricks and do other jobs. Bahishtan mentioned that the conditions for women were tough. 'Since my son Mutalli is a bachelor, where could he get a wife from to work with him at the kiln? So, his sister has to accompany him.' The contractors take liberties with *musalli* women and sexually abuse them. Girls cry helplessly. There is no bigger curse and helplessness than being a defaulter of a kiln contractor.

Nearly blind, Bahishtan narrated:

The plight of our family is very desperate. Sardara is licking the feet of another contractor to get his wife and children freed from the old one. We are starving, helpless and degraded. We do not have even a single penny. We have already borrowed a great deal of flour, pulses and vegetables from the local dealers. Although we deserve about Rs. 10,000 as our wages for three seasons, the contractor has instead put a claim of about Rs.18,000–25,000 on us. We have even tried to bribe the contractor and his scribe by presenting gifts to their families but all in vain. So, we don't expect mercy from any one but the skies.

Karmalli and his sister Sardaran took action. They sued the contractor in court. But rather than getting justice they ran from one court to another between Sargodha and Lahore for the entire decade of 1980s.

Bahishtan continued:

We don't sell ourselves to the contractors alone. Rather, there is another class which yearns for and loves to suck our blood and that is the class of landlords. What actually happens is that a *musalli* borrows money from the landlord and starts doing petty jobs for him. This landlord is no different from the contractor. He neither pays the affixed amount of Rs. 100 a month nor frees his *musalli* slaves. Instead, he is keen on making monthly deductions. It is again the *musalli* who suffers most from the mutual rivalries of the landlords. Like the contractors, the landlords also do not spare the honour of *musalli* women. It is very hard for a *musalli* woman to keep her chastity. Currently, there are about a hundred houses where *musallis* live in Miana Gondal but their locality looks deserted. This is because they are either living with the landlords or the contractors.

FROM SLAVERY TO DEEPER SLAVERY: PHASE TWO

The story of Bahishtan's sufferings does not end here. The landlord also uses *musallis* as bait in his various crimes and rivalries. He fires his guns by placing it on their shoulders. Most acts of murder and theft are not done by *musallis* at their free will. It is done at the instance of a landlord to settle scores with his rival. They simply cannot disobey any order of their master. The landlords commit crime and the report is entered against a *musalli*. At times, he also lays down his life for the landlords' protection.

Musalli are Born Borrowers

It was the unanimous voice of my village that *musallis* were born borrowers and spendthrifts. Everybody told me one or another tale to prove that they were spendthrifts and, therefore, never going to recover from their debts. The villagers believe *musallis* have been living like slaves, generation after generation due to their extravagant spending.

Musallis' Defense

I returned to Miana Gondal in March 1998. A couple of new streets and new houses had been erected near the *musallis*' locality. The original habitat of the *musallis*, a centuries' old pond adjacent to it and the surrounding fields had disappeared. While wandering through those streets, I found remnants of a few *musalli* houses. The house of Bahishtan had vanished altogether. However, in a shabby single-room, I met her youngest son Mutalli, who told me that the kiln-owners had left Karmalli and his family at Gujrat. There he is looking after the cattle of the landlord, Shana. Usually, the monthly salary for such a job varies between Rs. 200–400. As Shana has only two or three cattle Karmalli is paid Rs. 300 for a month. His other brother Sardara works at a brick kiln in Bhalwal along with his family. His sons work as wage labourers in Karachi.

'The *musalli* settlement in Miana Gondal is deserted', says Mutalli, 'because most of the people have gone to the kilns for labour. Their houses are locked up. They return on 30 June every year.' There was an elder *musalli* known as Mirza, whose son Bashir works at a veterinary hospital in the nearby town of Bar Musa. The second son looks after the cattle of the *maliks* whereas the third son works at a brick kiln in Mirpur. The work schedule at the kilns is disciplined and regular. The *musallis* go to work at the kilns in the beginning of February and return to their homes by 30 June. After a period of one month or six weeks, when the rains are over, they get back to work.

If the rainy season stretches a little longer than usual, they are deprived of their wages for that period. So they need to borrow more money. At times, the owners continue their work which saves *musallis* from the seasonal loss.

A worker at the kiln, known as a moulder, has a strange system of paying back the loan. If he is under debt for Rs. 10,000, half of his monthly wages would be deducted as a premium of his pay-back. Naturally, it is very hard for a worker to make both ends meet with the remaining salary. He, therefore, borrows more money.

At times, the owners and their scribes increase the borrowed amount in the register. The scribe even cheats the moulder while counting bricks. He writes 400–50 bricks instead of the actual figure of 500, thus depriving the moulder of his rightful earnings. However, the advantage of working at a kiln is that wages are based on quantity of bricks, so if the worker takes a day off, he can make up for it by working harder on another day. But if you work for a landlord, especially managing his cattle, the worker is busy all the time. He is expected to feed the animals even as late as midnight and get up again with the break of dawn.

Mutalli mentions that his job at the hospital is precarious, as he was appointed during the tenure of Benazir Bhutto. The Pervez Musharraf government is now aiming at dismissing all such appointees. Bahishtan's son Mutalli and his brothers do not respect their sister Sardaran because they feel she has dishonoured the family by working for the kiln-owner. When I reminded them that she had worked and suffered for her brothers, Mutalli said:

> There was no need for all that on her part. We would have certainly repaid her one day. Now she has been licking dust somewhere in Lahore. Neither do we know anything about her, nor are we interested in knowing anything about her. Her daughter also must have grown up by now. Karmalli has three sons and a daughter who are all married. Sardara has four sons. From our mother's side, we also have two stepbrothers, Sohbati and Salabti who reside near the mosque of *musallis.*

The *musallis* have their own separate mosque made of mud, a proof enough to show that it belonged to the poor. Now they have spent a lot to beautify and furnish it. One of the Khwaja family members claims that the landlords have renovated the mosque, but *musallis* refute this claim. Karmalli confirms: 'This is our mosque and we have spent on it.' The interesting thing about it is that its religious head is a barber who leads the prayers and teaches the children. The boys and girls can get education here up to the third standard.

Mutalli tells us that schools for both boys and girls has been upgraded up to the matric level in Miana Gondal. The girls' school is located in the street of the Samors (a local caste) and the main school for boys is situated near the village police station.

CLASS CONFLICT AND CHANGE

The essence of Mutalli's conversation was that *musallis* have lost their identity of caste and profession. Exploited first by the Khwajas, they then fell prey to the cruelties and inhuman treatment of the landlords and brick kiln owners. Their homes are deserted. What is symbolic of their life is that the stove is lit in their homes for only two months of the year. Their children have no permanent place to live. They are deprived of education, health and self-respect.

The houses of Khwajas have also disappeared from Miana Gondal. My family, descendants of Mian Fazil Karim, is slowly leaving the village. Two of his sons, who were my uncles, have passed away. His third son, Khwaja Muhammad Sharif, my father, died in November 1999. My father and my brothers lived in Lahore. My other relatives live in Mandi Bahauddin, Bhera and Dubai. The family home built by my grandfather is rapidly turning into a ruin.

Ghulam, 100 years old, is the only survivor to describe the nature of class and class conflict in this village, a representative of rural Pakistan. When I met him in April 1998, he started his story with: 'Your grandfather, you see, when he came to Miana Gondal. . . .'

The story depicted in these pages requests a detailed study and analysis. The landlords and owners of the kilns are reaping the benefit of the 'money-crop' (cash nexus) sown by my grandfather. The new generation of the Khwajas, mostly settled in the towns, continues to exploit the common man and the lower castes and classes for their wealth. The seriousness of the situation calls for immediate attention from the government as well as the non-governmental organizations.

RECOMMENDATIONS

The following recommendations are worth noting:

- The affairs of the landlords and the kiln owners must be thoroughly probed into.
- The *musallis* must immediately be liberated from the slavery of the 'dark ages'.
- They should have the protection of labour laws. For example, eight hours of work a day and enough wages to lead a decent life without 'selling' their wives, sisters, daughters and sons.
- Sexual exploitation of women must be eliminated through the implementation of criminal law.
- Since the *musallis* have been living as slaves for many years, they have in all probability re-paid their debts. Therefore they should be absolved of the current debts. Bonded labour is illegal and must be eliminated.
- Besides the government, it must be obligatory for the workers' trade unions, journalists and social welfare organizations to reveal the facts in this matter and help the *musallis* obtain respectable work that will pay for their daily needs.
- Civil society should focus on this crisis. An investigation must be made to look into the emergence of bonded labour and its current status.

INTERVIEWS

(1) Ghulam Lohar, Miana Gondal.
(2) (Late) Bahishtan Musallan.
(3) Sardaran Musallan.
(4) Sardar Musalli.
(5) Mutalli Musalli.
(6) (Late) Khawaja Muhammad Sharif, Lahore.
(7) Khawaja Muhammad Yusaf s/o Khawaja Fazal Ahmad, Miana Gondal.

REFERENCES

Bayly, C.A. (1983), *Rulers, Townsmen, and Bazaars: North Indian Society in the Age of British Expansion, 1770–1870*, Cambridge: Cambridge University Press.

Census 1981, Gujrat District.

Choudhry, Devendra K. (1982), 'Initiative, Enterpreneurship and Occupational Mobility in an Indian Caste: A Study of the Khatris of the Punjab (1881–1914)', *Proceedings of the Punjab History Conference*, vol. XVI, March, Patiala: Punjabi University,

Datar, Kiran (1985), 'The Traders as Administrators: The Khatris of Punjab', *Proceedings of the Punjab History Conference*, vol. XIX, March, Patiala: Punjabi University.

Qaisar, A.J. (1974), 'Role of Brokers in Medieval India', *Indian History Review*, vol. 1, no. 2.

CHAPTER 2

Impact of Colonial Capitalism on the Peasantry in West Punjab

Lubna Saif

INTRODUCTION

This paper deals with the impact of colonial capitalism[1] on the peasantry of Punjab in the theoretical framework of 'underdevelopment or dependency paradigm'. Colonial capitalism promoted underdevelopment in Punjab by redefining the relationship of power and property which created an 'elite' social class—a product of colonial capitalism. Relationship between power and property was most vital in Punjab, where more than 80 per cent of the population was engaged in agriculture. Agricultural colonization of Punjab was linked to the recruitment of the Indian Army in the aftermath of the 1857 Revolt. The focus of colonial policies was on the arid western parts of the Punjab, which became the main recruiting region for the Indian Army.

Western Punjab played the most vital role in strengthening the colonial rule by attaining the status of the 'Sword Arm of the Raj'. Agricultural colonization of western Punjab, considered by the British administrators as their proud achievement, was aimed at the transformation of a backward economy. It turned out to be a classic case of 'colonial development', rather underdevelopment (Ali, 1988). The agricultural colonization linked to the recruitment of the Indian Army brought rapid and extensive economic growth at the cost of the underdevelopment of the province. Consequently, Punjab underwent a dramatic change and took a shape which was essential for the promotion of colonial capitalism. The colonial capitalism completely distorted the pre-colonial structures and introduced the worst kind of feudalism in the disguise of agricultural reforms and economic development.

The underdevelopment paradigm uses dependency theory, according to which the interference of metropolitan economic connections, cultural influences and technology in the Third World[2] promotes underdevelopment rather than progress (Amin, 1973, 1976; Griffin, 1978; Franke and Chasin, 1980; Escobar, 1985; Rehnema, 1986). For our purpose of analysis of the causes of underdevelopment of peasantry in Punjab, dependency paradigm is the most appropriate tool offering many variations within the paradigm. Dependency theory explains the interdependent nature of the world capitalist system in its historical dimension, focusing on the total network of social relations as they evolve in different contexts over time. Since its crystallization in the 1960s and 1970s, though it went through a phase of disarray, the dependency paradigm is still the most powerful framework for the analysis of Third World underdevelopment. It also holds the potential of explaining the emerging wave of 'globalization' and its impact on the future of developing nation states.

The theoretical model proposed here suggests that in analysing the underdevelopment of 'national economies', it is important to understand the historical process of unequal development of the world, which began in the sixteenth century with the formation of a capitalist world economy and concluded in the post-Cold War era in today's 'globalization' or the 'new world order'.

THE UNDERDEVELOPMENT PARADIGM

Besides Marxism, there are two theoretical paradigms which explain the phenomenon of underdevelopment in the Third World, namely the development or modernization paradigm and the dependency paradigm. With the disintegration of the Soviet Union in 1991, the Marxist school is in a disadvantaged position to offer an explanation for the Third World underdevelopment. The modernization paradigm is narrowly constructed with its focus on the dynamics of change in a national society. It presents a culturalist thesis, in which a set of stereotypical notions of 'character' and 'personality' are found. According to these notions, colonial people were 'primitive' and 'barbarous', untouched by civilization until they came in contact with colonialism. Two schools of thought, evolutionism and diffusionism, representing the broader 'functional school' propagate this thesis. Most notable amongst the authors of the functional school are Gabriel Almond (1960), Joseph La Palombara (1963), Leonard Binder (1961) and Samuel P. Huntington (1968). All these distinguished political scientists were members of the 10-member Committee on Comparative Politics constituted in 1954, with the task to analyse and recommend development strategies for the newly liberated states in Africa and Asia. The theme of their analysis was in terms of 'crises of development', 'national integration', 'moderation', and so on. They advanced the pro-establishment school of development.

Both the evolutionists and the diffusionists view societies of the Third World as 'traditional' and 'static' before their contact with the West, lacking any 'endogenous dynamic for development and progress'. Without discussing the destructive role of colonial impact, they see colonialism as a modernizing force responsible for the transformation of a traditional society into a modern one. Discussing the 'self-legitimating ideas of colonial domination', K.S. Singh shows that there is not a 'hint or acknowledgement of the destructive role of the colonial impact in India' (Singh, 1978: 1221; also Chandra, 1980). The writers of the modernization paradigm reject any connection between the poverty of the underdeveloping world and the wealth of the developed world and they do not address the question of colonialism directly. They present a concept/model in which the villain is the 'non-Western world' and its 'cultural strains', which should wait for deliverance through the 'diffusion of world culture'.

The myth of the 'modernization or development paradigm' that poor are poor because they have not caught up with the modernization and industrialization is being challenged by writers of the underdevelopment or dependency paradigm, who view the historical appearance of colonialism and the emergence of Western capitalism as primary factors accounting for the existence of poverty in the world. There are various shades of 'dependency paradigm'. The most representative literature of the 'dependency paradigm' comes from Andre G. Frank (1968, 1972), Anthony Smith (1978), Peter T. Bauer and Basil S. Yamey (1978), Theotonio dos Santos (1984), Samuel Valenauela and Arturo Valenzuela (1984). They argue that the rich countries have become industrialized by exploiting the resources of their former colonies, and although political colonialism ended in the two decades following Second World War, the dependency of colonial relationship persisted even after the independence of colonial states. The neo-colonialism has created a world market which is based upon continuing unequal trade relations between the rich and the poor world (Wallerstein, 1984, 1974; Murdoch, 1980; Adelmann and Morris, 1973; Griffin, 1978). The poverty in the world owes its existence to 'this world economic system' (Griffin, 1978; Frank and Chasin, 1980; Escobar, 1985; Rehnema, 1986). This results in the division of the world between industrial, advanced or 'central' countries, and underdeveloped, backward or 'peripheral' countries. The 'centre' is seen as capable of dynamic development sensitive to the internal needs of the whole, and as the main beneficiary of the global links. On the other hand, the 'periphery' is viewed as having a reflex type of development; one, which is both, constrained by its incorporation into the global system and which results from its adaptation to the requirements of the expansion of the centre.

Writers of the dependency theory argue that the development of a national or regional unit can only be understood in connection with its historical infusion into the worldwide political-economic system that evolved with the wave of European colonization of the world during the fifteenth century. This global system is characterized by the disproportionate but connected development of its different components. Frank suggests that the causes of underdeveloped status are linked to the historical causes of development in the industrialized nations (Frank, 1972). According to the authors of the dependency theory, underdevelopment and development are two sides of the same coin, both are historically simultaneous, both are linked functionally and both interact and condition each other.

IMPACT OF COLONIALISM ON LANDHOLDING AND AUTHORITY

With the advent of the British, the pre-colonial economy of India underwent a transformation. Colonial rulers brought with them established eighteenth-century ideas of absolute property and the prime value of a landowning gentry or aristocracy as the backbone of a nation. It was natural for them to search in India for the 'rightful' proprietors of the land and when they did not find any, they created them by changing the 'rights to rule' the land to 'rights to own' the land (Habib, 1984). By doing so the British legalized the private right to absolute land ownership. They repeated this practice in different ways in different parts of India with similar consequences: the cultivators had to give away their customary rights, while most of the land ownership concentrated in the hands of the upper classes. The primary motive to create a landowning class was to provide 'leadership' with whom the British could bargain (Sayeed, 1980; Alavi, 1983; Gardezi, 1983).

In a predominantly agrarian society such as Pakistan, the ownership of land is a key issue in development. Much of the country and in particular its most productive areas are still characterized by highly inequitable forms of land tenure. Despite three land reforms, a large number of cultivators are sharecroppers on land owned by a few, in most cases absentee landowners. As Mahbub ul Haq notes: 'Pakistan's capitalistic system is still one of the most primitive in the world. It is a system in which economic feudalism prevails' (Haq, 1973). This 'capitalistic system characterized by economic feudalism' is the product of colonialism that introduced capitalism in agriculture with the recognition of the institution of private property, and initiating commercial farming. This brought major changes in the political economy of India and particularly in the Punjab province. Hamza Alavi observes:

> The colonial impact brought a specific, colonial type of 'bourgeois revolution' in the colonies, establishing a structure of specifically colonial capitalism. Given that structure, various internal developments follow in the process of the creation of a colonial economy. (Alavi, 1980: 397)

Regions that experienced an enormous volume of contact with developed countries in the past are today's most depressed regions. Frank observed that this contradicts the thesis that underdevelopment results from isolation and pre-capitalistic institutions (Frank, 1966). The colonial experience affected every aspect of a colonial society. Land distribution, land tenure, agricultural practices, social and legal traditions, political authority, the national economic structure, regional and international economic relations were so fundamentally distorted that the damage continues to affect the development of ex-colonies to this day. According to Gallagher et al.:

> The British wanted to pull resources out of India, not to put their own into India. Therefore, the administrative and military systems had to pay for itself with Indian revenues. . . . The chief source of Indian revenue lay in land and it had to be collected from millions of payers. . . . It was in the administration of localities that the vital economies in ruling had to be made. There, governance had to be pursued by simpler arrangements. . . . By enlisting the co-operation of zamindars, *mirasidars, talukdars,* and urban *rais.* (Gallagher et al., 1973: 8)

The social and political consequences of this colonial economy were the creation of 'elite' or dominant local interests given the nature of class arrangements emerging from the characteristics of 'peripheral economies' (Murdoch, 1980). The explanation for the continuation of underdevelopment in traditional societies like Pakistan is to be sought in the process through which the colonial power established local economic and political elite in the colony (Gardezi, 1983; Alavi, 1980; Sayeed, 1980), especially in the frontier provinces of north-west India, which had a strategic importance for the defense of British India. The focus of inquiry is the province of Punjab, which became the stronghold of the British colonialism by contributing the most loyal recruits for the Indian Army and earned the reputation of the 'Sword Arm of the Raj'. This paper explores how Punjab acquired the status of the 'Sword Arm of the Raj' and served the colonial interests as the most 'loyal province'.

With the advent of colonial rule, Indian society underwent profound changes, especially in terms of 'the structure of relationship between the cultivator and the social relations of production' (Alavi,

1980: 370). Three major colonial institutions can be identified, which transformed these social relations of production—the institution of property, civil bureaucracy and the legal system. These three institutions were interrelated; the establishment of the colonial legal system was linked to the question of ownership of property, and the institution of civil bureaucracy was meant to translate these ideals into action. We cannot evaluate the impact of the 'colonial capitalism' without understanding pre-colonial structures of agrarian relations and land revenue.

PRE-COLONIAL STRUCTURES OF AGRARIAN RELATIONS AND LAND REVENUE

In northern India, the concept of absolute private ownership was not present before the arrival of British in the eighteenth century (Gardezi, 1983; Sayeed, 1980; Raychaudhuri and Habib, 1984; Alavi, 1980). Rights to land, not the land itself, belonged to the group: the village, extended family, or the lineage group (Sayeed, 1980; Naseem, 1980). Occupancy rights for cultivation were in farmers' hands. The farmers which were not allowed to transfer or alienate land to someone else; they could only hand down occupancy rights to their heirs (Naseem, 1980). 'However, it was possible to mortgage or sell the occupancy right to members of the community, though there was a custom of pre-emption, which required that the decision to transfer be not made individually' (Habib, 1984). Habib writes:

> There was no question of free alienation—the right to abandon or dispose of land as its holder might choose which is an essential feature of modern proprietary right. . . . He could not leave it or refuse to cultivate it. (quoted in Naseem, 1980: 14)

To provide the bases for colonial power in India, British colonial rulers exploited the Mughal mode of production. The Mughal mode of production was different from colonial capitalism because it had no concept of 'private ownership'. Habib notes, 'the ownership was by and large confined to particular rights over land and, not . . . as an alienable and unrestricted claim over land' (Habib, 1984:

176). Alavi argues that the '*zamindari* rights of Mughal period that could be bought and sold' cannot be considered as property rights (Alavi, 1980: 371). Gardezi observes that, 'the Mughals, in particular, introduced measures to prevent their official aristocracy from developing strong property rights in land' (Gardezi, 1983: 24). He notes that strong property rights was the creation of British colonial policy (ibid.). Analysing British colonial policies, Imran Ali writes: 'they encouraged individualization in property rights, which was a marked shift from the collective ownership by village communities . . . existed in the pre-British period' (Ali, 1988: 4). He observes that with the process of individualization, 'probably for the first time' the prospect of alienation of land rights was created (ibid.).

The term *zamindar* is a Persian compound, which means keeper or holder of land. 'The term began to be used more and more often from Akbar's time onward for any person with any hereditary claim to a direct share in the peasant's produce' (Habib, 1984: 176). During the seventeenth century, the term *zamindari* replaced all other local terms for agrarian rights of any kind (ibid.). In the larger part of Mughal India, the *zamindar* was expected to collect the tax from the primary assessees (cultivators), in return simply for an allowance of one-tenth, given either in cash or in allotment of revenue-free land. The important fact to note is that the 'Mughal land revenue' or *mal* was not rent or even a land tax like in any feudal economy. It was sharing of a crop—*ghala batai* or tax on the surplus or 'a claim on behalf of the state to a share of a crop' (ibid.: 235). *Mal* had nothing in common with the 'land revenue' of British settlements, which was instituted as a standard burden fixed on a particular area of land, without taking into account the produce of the land (ibid.).

The Mughal administration was based on the *mansabdari* system, in which administrators were ranked into different *mansabs* in accordance with the services and duties assigned (Habib, 1963; Spear, 1965). Though the *mansab* was not inheritable, however, in most of the cases sons or relations of higher *mansab*-holders were given due consideration. It was a tenure job, an uncertain tenure dependant upon the emperor's will (Habib, 1984). In addition to paid salaries, the *mansabdars* were assigned *jagirs* or estates; *jagirs* were transferred on a constant basis to weaken the *mansabdar's* ties with the land and its cultivators (Spear, 1965). By imparting temporary character to the *jagir*, the control of the emperor was further strengthened (Habib, 1984). Habib writes:

> The *mansab*-holder was entitled to a *jagir*, but not to a particular tract of land in *jagir*, and not the same land year after year. The *mansabs* were revised from time to time to award promotion or demotion. . . . Similarly, there were transfers of officials from one province to another; in each such case, a block of territory had to be carved out for the official in the new province . . . the result was that no one could be sure of how long he would remain in possession of a particular area. (ibid.: 243)

The Mughals introduced an elaborate and complex system of revenue collection, in which 'alternative and overlapping institutions for revenue extraction—*khalisa*, *jagir*, *zamindari*, and *raiyati* villages— . . . were not neatly segregated one from the other'.[3] This system rested profoundly on the *zamindars*, whose rights and powers were reinforced, originally derived from outside the imperial system (Habib, 1984). These *zamindari* rights, 'heritable, saleable and actually mortgaged', were not uniform, and created a heterogeneous body of *zamindars*. As noted by Habib: 'some *zamindars* paying tribute, others holding their land as *jagirs*, while still others acquired non-hereditary rights by royal order'(ibid.: 177).

With the development of revenue farming, 'powerful *zamindars*' were able to purchase *zamindari* rights of other *zamindars* and a new concept of *taluqdari* came into vogue defined 'as a newly purchased *zamindari* and a tenure which entitled its holder to engage on behalf of other *zamindars*' (Siddiqui in Raychaudhuri and Habib, 1984: 177). *Taluqdari* was common in two Mughal provinces, Bengal and Awadh but it did not exist in Punjab. The sale price of *zamindari* rights was in relation to the annual land revenue. For instance, the sale price of *zamindari* in five villages in the 1670s and 1680s in the province of Awadh amounted to nearly double the annual land revenue.[4] Habib observes that the most important feature of the *zamindari* right was its uneven development. Within the same district, some villages would fall into the category of *zamindari* villages and some villages would be (peasant-held) *raiyati* villages. The state did not favour any particular type of villages—there are numerous instances of establishing fresh villages without any intermediation and at the same time supplementing *zamindari* villages by creating *zamindari* villages (Habib, 1984).

Three basic institutions, the family, the caste and the village community regulated pre-colonial Indian society (Sovani, 1954). Customary laws and traditional rules governed these institutions outside the preview of the state. *Panchayat* or the council of elders was the arbitrator. The vastness of the area never allowed any central state to interfere in the social organization of people (Hassan, 1969). Because of their isolation, few villages were linked with the central power. The following description of the village community provided by Habib illustrates this point:

> The Indian village presented the appearance of a closed, custom-based social and economic unit. The close settlement of peasant households and the needs for peasant migrants to move in a body for better protection furnished the basis for a collective organization of peasants, within the framework of clan and caste, the Indian Village Community . . . to a certain degree an Indian village was a stable economic unit, self-sufficient in respect to its own consumption needs. (Habib, 1969: 37)

PRE-COLONIAL AGRARIAN STRUCTURES OF PUNJAB

Punjab was a landlocked region in the north-west of India, situated at the traditional route of trading caravans and invading armies coming from the Central Asia through the passes of Afghanistan. It was always the first province of the subcontinent India to be annexed—a gateway to India that had been a part of many Central Asian empires in various historical periods and subject to numerous invasions. The Punjab has been described as a passage of human movement from Central Asia to the northern India, a 'cultural and political route zone' (Cohen, 1971: 25). The majority people of Punjab belonged to the Central Asian nomadic tribes who accompanied the invading armies and settled in its plains.

Its location and an extensive river system were the two main influencing factors in the making of Punjab's history. The province derived its name from its five major rivers, Punjab meaning the 'land of five rivers'. It had been cradle of many civilizations; Harrapan and Gandhara civilizations adorn its past. Punjab is the home of India's major religions including, Hinduism, Buddhism, Islam and Sikhism.

In the Mughal Punjab, two types of *zamindars* could be identified: (i) intermediary *zamindars*, and (ii) primary *zamindars* (Hassan, 1969). The former were responsible for revenue collection in *zamindari* villages and possessed a right to a certain share of the produce of land, which they did not cultivate themselves. The latter were peasants of *zamindari* and *raiyati* villages who were given the proprietary rights to cultivate the land themselves or with the help of hired labour.[5] In Punjab, *zamindari* and *raiyati* villages were more familiar and the class of autonomous chieftains found in other Mughal provinces was absent. In *zamindari* villages, *zamindars* were responsible for the collection of revenue, while in *raiyati* villages, the peasant community enjoyed a relative degree of autonomy from the *zamindar* (Alavi, 1980), and the revenue was collected by *muqaddams* (the village headmen). Gardezi notes that the *muqaddam* was a 'more permanent figure in the social structure of the village . . . he often cultivated his own land and was directly responsible for the collection of the revenue from the villagers' (Gardezi, 1983: 24). *Raiyati* villages were coparcenaries, the lands and wells being divided amongst cultivators on the basis of fixed, customary shares. Those tenant cultivators who enjoyed hereditary occupancy rights and paid the revenue themselves were counted primary *zamindars* too.[6] During the Sikh rule and the British Raj, the *muqaddam* was replaced by the *chaudhary* and the *lambardar*. Though the functionary remained the same, the structural changes inducted completely altered its function, particularly in the British Raj, when the *lambardar*—paid local agent of the Raj—replaced the traditional institution of the 'headman'. The Land Administrative Manual prepared by the early British administrators stated:

> The affairs of the brotherhood were formally managed by an informal village council or *Punchayat*. But this body was too numerous and loosely constructed to fittingly represent the community in its dealings with the Government officials. A few of its leading members were, therefore, selected as

headmen or *lambardars*, and the appointment of headmen naturally came to be confined to particular families. From a revenue point of view the most important function of the headman is to collect the revenue from the coparceners and pay it into the treasury. The social position assigned to the *lambardars* and the action of our courts stripped the *Punchayat* of its influence, and practically it has ceased to exist. (Moudie, 1908: 62)

In the eighteenth century, with the collapse of the Mughal Empire, the provinces became autonomous and the Mughal viceroys established their independent rule. Though the decline of central authority led the way for regional autonomy, the Mughal political and social structures continued, especially in the provinces of Awadh, Bengal and Hyderabad. In Punjab this social and political continuity was interrupted following two reasons. First, there was no class of local chiefs such as the *taluqdars* of Awadh or the *zamindars* of Bengal (Baden-Powell, 1892), who could establish regional autonomy. Second, the province had become subject to various Afghan invasions. This caused complete social disorder and the Punjab became a hunting ground for all kinds of warlords and its political economy was in a constant state of disintegration. By the last half of the eighteenth century, on the eve of the Sikh ascendancy, Punjab was in total disarray. The Sikh ascendancy began with the Sikh peasantry's revolt against Mughal intermediary *zamindars*' agrarian policies of extraction of surplus and concluded in the organization of *misls* or armed bands. The Sikhs had a history of conflicts with the Mughals, their eighth guru Tegh Bahadur was executed by Aurangzeb's order in 1675. The latter's religious intolerance forced the last Sikh guru, Gobind Singh to transform Sikhs from a passive sect into a militant brotherhood of the Khalsa. With the collapse of Mughal authority, the Sikhs started to gain control of the plains of Punjab, and 'their armed confederacies took advantage of the power vacuum to carve out petty state' (Talbot, 1988: 32). By 1823, Ranjit Singh had completely established his sovereignty by replacing the Mughal aristocracy and subjugating all other *misls*. Major writes:

The Kingdom of Lahore, proclaimed by Maharaja Ranjit Singh in 1801, was a military patronage state, initially created by Ranjit Singh's subjugation of the powerful Sikh *misls* of central and eastern Punjab. (Major, 1991: 54)

The Sikh state was a military state and all powers belonged to Ranjit Singh. Though Sikh armed bands or *misls* revolted against Mughal intermediary *zamindars*, Ranjit Singh continued the Mughal system of revenue and the practice of granting *jagirs* to the governors in lieu of services. However, the layer of intermediary *zamindars* was abolished and the State extracted surplus from the peasantry in exorbitant proportions through a system of 'highest bidder'.[7] After Ranjit Singh's death, the Sikh state collapsed[8] and soon after his death, the British adopted an aggressive policy in an attempt to subjugate the Punjab. Between 1803 and 1849, most of the territories of the provinces were integrated into the British Punjab.

ESTABLISHMENT OF COLONIAL CAPITALISM IN PUNJAB

According to the *Census of 1868*, the population of Punjab was around 9.5 million, excluding the princely states. Muslims constituted 63 per cent, Hindus 22 per cent and Sikhs about 8 per cent (*Punjab Census*, 1870). Punjab was the second last province to be annexed by the British. At the time of its annexation, the territorial domains of the province included undivided Punjab in the west with present-day Haryana, Himachal Pradesh, Delhi and the north-west districts. On 25 October 1901, the five districts in the north-west frontier were carved out to form the North-West Frontier Province (NWFP), comprising territories, which lay west of the river Indus. Though the British did not follow the traditional route, Punjab's location—its proximity to the erstwhile USSR—and its extensive river system were determining factors in shaping the colonial development policies (Talbot, 1988).

Ranging from Delhi to Attock, Punjab was characterized by religious and geographical diversity. The crucial factor for the geographical diversity was the differing availability across the region of water for cultivation and natural vegetation. Three hydraulic zones could be identified: (i) the moist zone, (ii) the marginal zone and (iii) the arid zone. At the time of the census of 1911, the following natural divisions as shown in Table 2.1 were recognized.

TABLE 2.1: NATURAL DIVISIONS OF PUNJAB

Name	Total areas (in sq. miles)	Total population	Density per sq. mile	
			Total area	Cultivated area
Himalayas	22,050	1,724,480	78	965
Sub-Himalayan	19,045	5,805,081	305	612
Indo-Gangetic Plain	8,525	11,027,490	286	435
North-western dry area	56,710	5,630,699	99	432

Source: M.S. Leigh, *The Punjab and the War*, Lahore: Government Printing Punjab, 1922, p. 1, PA.

The *Census of 1921* shows that more than 55 per cent of the population was Muslim and concentrated in the north-western dry areas, consisting of the marginal and arid zones, while the Sikhs, an important minority, were based in the central moist zone of the province in the predominantly Hindu districts (see Table 2.2).

TABLE 2.2: COMMUNAL RATIO OF THE POPULATION OF PUNJAB IN 1921

Muslims (%)	Hindus (%)	Sikhs (%)
55.33	33.58	11.00

Source: Government of India, *Census of 1921*, Calcutta: Government Central Printing, 1923, PA.

In the following discussion, an analysis of the process of establishing colonial capitalism and its impact on the social, political and economic structures of Punjabi society is presented.

PUNJAB AS THE SWORD-ARM OF RAJ AND CREATION OF LANDED ELITE

Punjab's rise as the 'sword-arm of the British Raj' and colonial capitalism are inseparable. The character of economic structure developed under the colonial capitalism in India, including Punjab had three features. One, 'the institution of private property in land'; two, 'the growth of merchant capital in Punjab'; and three, 'the establishment of a cash nexus as the primary form of surplus extraction by the colonial state' (Pasha, 1998: 108). This economic structure resulted in the marginalization of the peasantry and the concentration of land in a few hands, which ultimately pushed peasants into the colonial army. The British Imperial interests were dependent upon an Indian Army, which not only stood in reserve to maintain law and order throughout the subcontinent, but fortified British control of the Middle East and the Indian Ocean (Talbot, 1988). At the outbreak of the First World War, Indian Army was the largest volunteer army in the world's history, and 'consisted largely of officers and men drawn from one province, the Punjab' (ibid.: 41). The period 1875–1914 witnessed a complete alteration of the composition of the Indian Army. Talbot observes:

> At the beginning of this period Punjabi troops accounted for just a third of its total strength and the army's main recruiting areas still lay in Bengal, Madras and Bombay. By the end, however, three-fifths of the troops came from the Punjab and the Army had made its home there. (ibid.: 41)

The support of local allies was crucial to the colonial control of the Punjab for its immense value to the British strategic interests and its reputation for political loyalty to the Raj (ibid.). The colonial state used 'the land settlement policies', especially in the Canal Colonies for the fulfilment of military needs and provided rural bases of military recruitment patterns (Ali, 1988). The penetrating process of the military into the 'seminal source of societal authority, the control and possession of agricultural land', cannot be seen in the 'context of modernization', rather it 'would be more appropriate to draw parallel with feudalism' (ibid.). Subordinating development goals to its political and military imperatives, the colonial state promoted the social elite, the members of which became the major beneficiaries of agricultural colonization (ibid.).

The cooperation between the large landowners—a class of Punjab chieftains created by colonialism—and the British colonial state was primarily motivated by the Punjab's emergence as the sword-arm of India in the aftermath of 1857 Revolt. Colonial administrative policies in the 1860s and early 1870s generated this class of Punjabi chieftains that not only dominated the political and economic life of the province in the years of the Raj but also continued to play the most important role in the creation of the post-colonial state of Pakistan. For our purpose of inquiry, the focus is on the creation of the class of Muslim chieftains of western Punjab, a region that became the 'flower bed' for the recruitment of Indian soldiers.[9] This development was rooted in some rural families' response to the 1857 Revolt (Talbot, 1990; Page, 1987). David Page wrote that 'the Muslims of West Punjab had played an important role in the annexation of the province, and both the Muslims and the Sikhs provided the forces, which enabled the government to put down the Mutiny of 1857' (Page, 1987: 49). These Muslim rural families were the ones who provided military services to the Sikhs and served the Sikh *durbar* in various administrative capacities, and had been given *jagirs* (Gilmartin, 1988). Most of these families were of peasant origin and attained a prominent status by supporting and in some cases joining winning *misls.*[10] Their real fortunate moment came in 1857 when they joined and helped the British officers in the 'dark and thick clouds' of the 'Revolt of the Indian Army'. The British recognized the heads of these families as 'tribal leaders', their local allies without whose support a strong loyal Punjab geared to bear the burden of Indian Army could not be built. These 'tribal leaders' were promoted as 'chieftains', representatives of the landed class, since 'the British found few established 'tribal' leaders in Punjab as they began to construct their rural administration' (ibid.: 18).

In absence of any 'traces of former leadership', structuring a reliable rural administration was a great challenge for the early British administrators. They constructed this foundation by re-designing the Punjabi rural structures and creating a 'tribal social organization' in which 'tribes' became the basis of authority (ibid.: 18–38). Establishing a landowning class based upon 'tribes' in Punjab is linked to the 'colonial recruitment policy of Indian army'. The Punjab government 'very heavily relied on this landowning class in the recruitment of soldiers for the Army' (Page, 1987: 49). The relationship between the Muslim landed leadership and the

Raj was determined by two factors: 'the first administrative and the second military' (ibid.). In the following discussion, we will examine how this Muslim landed leadership was created and sustained through a special set of administrative arrangements for the province of the Punjab.

The analysis of British colonial policies in north-west India reveals how the British in the quest of their military and strategic interests on the borderlands were dependent upon the support of certain tribal and landowning interests. Highlighting the contrast between British policies in north-west and other parts of India, Talbot argues that security rather than commercial considerations prompted the British administrative policies in Muslim north-west India. He writes:

> The British interest in harnessing the waters of Punjab and Sindh notwithstanding, their advance in Muslim north-west India was prompted by military rather than commercial considerations. . . . Their advance coincided with growing anxieties about threat to British India from Afghanistan and expanding Russian Empire in Central Asia. . . . Punjab was thus seen as a base for military operations, whilst the Frontier and Balochistan regions were buffer zones against a possible attack. (Talbot, 1990: 58)

Similarly, David Page observed that the 'fear of Russian expansion' and 'the location of the Punjab' created the dependence of Indian defense upon a strong Punjab (Page, 1987). Talbot argues that 'it is unlikely that the region would have assumed importance as a centre of colonial military recruitment, if it had not been near the Indian Army's main theatre of war in Afghanistan' (Talbot, 1988: 42). However, most of the British administrators traced this development in the turbulent history of the Punjab, and claimed that the Indian Army was built on the 'military prowess' and 'martial values' of the people of the Punjab (MacMunn, [1930] 1979). The theory of *martial races* was the official explanation for specialization of Punjab as the sword-arm of the Raj. The chief architect of the theory of martial races in India was Lord Roberts, the commander-in-chief of the Indian Army from 1885 to 1893.

Though the theory of martial races is full of inconsistencies, it is still being used to provide an explanation for the colonial recruitment policy. Surprisingly, in the post-colonial period, the military authorities find justification in propagating this theory for the over-representation of certain ethnic groups and the under-representation of others in the army (S.P. Cohen, 1984; 1983; 1980). There is a strong argument that Pakistan's 'overdeveloped' administrative and military institutions have their roots in the colonial legacy. Colonial rulers' concerns forced them to establish a 'security state in the north-west India'. Such a state could not afford political participation of the people, resulting in far less developed political institutions and its survival dependant upon the support of certain tribal and landowning interests (Cohen, 1984). This is evident in Sir Michael O'Dwyer's speech delivered to the Imperial Council in 1917 in which he explained that the Punjab government relied on two classes, the landed aristocracy and the mass of peasant proprietors, and according to him these two classes were not interested in political reforms.[11] Giving the details of the legislative and administrative measures to protect the landed interests, Sir Michael O'Dwyer very proudly declared that 'all these things . . . [were] done in the interests of our *zamindars* and especially of those tribes and classes which enlist[ed] so freely in the Indian Army' (ibid.). These 'tribes and classes' were classified by Sir Michael O'Dwyer as 'martial classes' (Chelmsford Papers).

Examining the structures created by the Raj, Talbot writes: 'the strongest alliance between the colonial state and the leading landowners existed in its Punjab bastion' (Talbot, 1990: 61). He pointed out that the 'long term influence of the British reliance on the collaboration of local rural intermediaries' not only 'jeopardized the achievement of Pakistan, but also threatened its post-colonial democratic development' (ibid.). The collaboration resulted in 'retarded political institutionalization' and introduced a culture of 'clientelism' (ibid.). These local rural intermediaries were created through a process started with the 'land settlement' policy, which resulted in the creation of homogeneous administrative units called *zails* and took its final shape through the Land Alienation Act of 1900, in the form of a system of indirect rule resting on the dominance of feudal and semi-feudal landowners and *pirs*, who were instrumental in the recruitment of Punjabi soldiers.[12] In the backward districts of western Punjab like Muzzafargarh, Multan, Sahiwal, Shapur and Jhang, many families of loyal *pirs* were converted into major landowners under the Land Alienation Act of 1900 by getting the status of agricultural tribes whose land could not be alienated.[13] By declaring them agricultural tribes, the British not only provided protection to these hereditary custodians but also made them *zaildars*, honorary magistrates and district board members (Sayeed, 1980). These hereditary custodians of shrines were also given maintenance grants subject to the good conduct of their wardens, in addition to 'landed gentry's grants' in the Canal Colonies.[14] 'The Punjab Government's recognition of the *Pirs* as part of the "landed gentry" had important political repercussions in establishing a unity of political interests between the landlords and the *Pirs*' (Talbot, 1988: 60). The formation of the Unionist Party in the early 1920s was based upon this close cooperation, which provided a strong support for colonial rule, which continued till the last days of the Raj.

Along with the *pirs* a class of big Muslim landowners emerged in west Punjab. As discussed earlier, these were the families that had proved their loyalty to the East India Company in the 1857 revolt. In Talbot's words, 'many rural Punjabi families embarked at this time on what were to prove lengthy and lucrative "loyalist" careers' (ibid.: 61). Around 47 Sikh and Muslim families were rewarded for their collaboration in 1857; they received titles like Khan Bahadur or Sardar Bahadur, military honours like the Order of British India or the Order of Merit, *Khillat* (a reward/robe of honour), *jagirs* and land grants—either a proprietary title or a landlord title—were conferred.[15] For example, Tiwana Maliks in Shahpur district were rewarded for their past services and bound to the new colonial regime by the creation of new *jagirs* (Ouselev and Davies, 1886). Those, who received land grants were also presented with *sandals* (deed), in which it was specified that the continuation of the grant was conditional on the good conduct of the grantee and his family.[16] On 26 May 1860, the Punjab Government announced the names of 25 'loyal chieftains' who were to be given special powers of an Assistant Commissioner and were called *Jagirdari* Magistrates. These were limited administrative powers, which enabled them to only decide those revenue and criminal cases whose value did not exceed Rs. 300. They were also made in-charge of the village constabulary within their estates and for these police duties, they were to receive *inams*. By 1865, the number of these *Jagirdari* Magistrates was increased to 38.[17] The principle of conferring limited judicial powers was extended to allow appointment of other 'chieftains' and 'influential' Punjabis as Honorary Magistrates. They were given the judicial powers to investigate and decide petty criminal cases that were in the jurisdiction of the regular courts in cities.[18] Later on, their powers were expanded to include certain civil cases also. Gradually the system of Honorary Magistrates was extended to the smaller district courts.[19]

The Court of Wards Administration was another important tool through which the colonial rulers reinforced the loyalty of the class of 'chieftains'. It was established to create and maintain estates for loyal families. The inheritance of these estates was dependent upon the evidence of loyalty to the colonial government. The law of primogeniture gave the right to British authorities to withhold recognition of certain politically undesirable inheritance arrangements. Through the Punjab Laws Act of 1872, the government was able to bring chiefly estates within the jurisdiction of the Courts of Wards, which gave deputy commissioners control over the affairs of these estates.[20] In 1895, there were 65 estates with an area of over 3,44,000 acres under the control of the Court of Wards Administration.[21] To strengthen the 'chieftain class' and attach it to the British rule, an education policy was designed. Aitchison College was established in Lahore in 1886 to provide education for the sons of leading landowners. Its admission was restricted to allow only sons of loyal chieftains. A large number of these Aitchisonians went on to have important political careers under the colonial patronage. The 'class of chieftains or big landowners' created by the Act, and through systems of honours, nominations to local *durbars* and district boards and appointments as honorary magistrates and *zaildars*, became the guardian of the British rule and with its alliance, Punjab became the major recruiting area of the Indian Army.

During the First World War, rural Punjab proved to be the most loyal province of the British Raj by responding to the government's campaign for enlisting recruits for the Indian Army to fight along with the Allied forces.[22] Three rural communities provided the lion share of soldiers, the Muslims of west Punjab, the Jutt Sikhs of central Punjab and the Hindu Jats of the Ambala division.

The five districts of Rawalpindi division (Rawalpindi, Jhelum, Attock, Gujrat and Shahpur) were amongst the eight most heavily recruited districts in the entire Punjab, the other two were Ludhiana and Amritsar, the two main Sikh recruiting areas.[23] 'The north-western districts especially were areas of concentrated recruitment of soldiers, and much of the activity of elite formation was based around the organized delivery of these recruits to the army'.[24] Recruitment was not voluntary. Under the oppressive government of O'Dwyer, the war efforts included the collaboration of *pirs* and big landowners for forced enlistment.[25] The other major reason for recruitment from the *barani* tracts of western Punjab was under-development and poverty created by the colonial capitalism, which pushed the poor peasant into the army, since there was no other opportunity of employment in this arid land which was dependant upon rain, and had a very low level of productivity.

Rural elite of Punjab were rewarded for its response to the 'government exhortations to enlist' for the War by the Punjab government in its proposals to the Montague–Chelmsford Report, in the form of distribution of urban and rural seats and its recommendations concerning franchise. According to the reform report, the council was to contain a minimum of 50 members, and 30 per cent members were to be nominated. Besides, 34 members were to be elected. The distribution of these elected seats displayed a 'strong bias' towards rural 'loyal' groups; 25 seats out of 35 were to be given to the rural areas. Of the rural seats, 10 were allocated to the Muslims, 6 to the Hindus, 5 to the large landowners and 4 to the Sikhs. The government also proposed a special treatment for the Punjabi electorate that had to be more restricted which made it the smallest one in India. Punjab was a 'security province' where large-scale political participation could not be allowed in the 'interest of the Raj'. Therefore, special qualifications were proposed for the franchise in Punjab. The most striking example to control the franchise was giving the right to vote to the officially appointed *lambardar* in each village, which made it impossible for any candidate to win without official patronage.[26] The allocation of five seats to the big *zamindars* also indicated the government's desire to strengthen the influence of its 'proven loyal elite', control the rural population and its presence in the council was meant to check the liberal educated urban representatives in control.[27] Similarly the government in its desire to secure an adequate representation for the army, proposed separate representation for the retired military men, not only by nomination but also through the rural electorate.[28] In response to the Franchise Committee Report, the recommendations of the Government of India and the decision of the Joint Select Committee, the Punjab government had to modify its few proposals and accept a larger council and a larger electorate under the Montague–Chelmsford constitution. However, the distribution of seats continued to favour the rural elite and in addition, its proposal for enfranchizing the Punjab soldier was accepted by the House of Commons, despite the opinion of the Joint Select Committee that 'this might mean that the soldiers in the Punjab would have a preponderating voice in elections.'[29] In the final constitution, the right to vote to the ex-sepoy was awarded, making the military vote the most substantial element in the rural electorate.[30] As a result of 1919 Reforms, the men who returned to the council after 1920 were those who were committed to the 'maintenance of the Punjab military machine' and they were facilitated to form the government. Consequently, the classes recruited to the army were the most privileged under the new constitution, getting maximum benefits.

As the chief representative of the military lobby, Sir Umar Hayat Khan, whose son became the last premier of colonial Punjab, had proposed to the Punjab government that military classes should be given their own representatives along with the landholders.[31] Sir Umar Hayat was speaking on behalf of the Punjab government. He was the representative of the landed class, created and sustained by the colonial policies to maintain Punjab's status as the 'sword arm of the Raj'. The Punjab government's reliance on this class was in show again when the Punjab Council was formed under the

Montague–Chelmsford (1919) Reforms. The large majority of Muslims who dominated the council in the following years belonged to the Punjab Muslim Association and was from the 'families of note' involved in recruitment during First World War. In Punjab, the Montague–Chelmsford Reforms provided an excellent opportunity to all the partners of colonial control to strengthen their bond of alliance. The division of seats in the Legislative Council on the basis of urban and rural constituencies, provision of vote to officially appointed *lambardars* and ex-sepoy consolidated the bargaining power of rural elite in favour of the Raj and retarded the process of nationalism.

> When a degree of democracy was introduced into the province, the servicemen and their comrades exerted a powerful influence . . . being granted land or government office as a reward of their loyalty, they came to dominate the restricted electorate. (Talbot, 1988: 46)

The emergence of the Punjab National Unionist Party in 1923, as representative of the military lobby belonging to the three communities—Muslims, Sikhs and Hindus—though predominantly Muslim, the body should be seen in this perspective that continued to rule Punjab till the last days of the Raj. Two leading personalities of the Unionist Party, Sir Sikandar Ḥayat Khan, the first premier and Sir Khizer Hayat Tiwana, the last premier, belonged to families who had a history of loyalty to the Raj, and were instrumental in making the Punjab the 'flower bed of Indian Army'. The Punjab National Unionist Party was borne as an alliance of the Punjab Muslim Association and the Punjab Zamindar Central Association representing the landed and military interests in the aftermath of the Reforms and continued to dominate Punjabi politics for almost a quarter of a century before the Partition. The Unionist ideology was consistent with dependence on a rural hierarchy.

The 1935 Act introduced a new element in Indian politics: provincial autonomy. It gave the Unionists a new weapon to consolidate their base in Punjab and keep it isolated from all-India politics. The Unionism did not allow political participation of people and growth of democratic institutions. A legacy Punjab inherited and continued after the Independence.

The landed elite created by the Raj was quickly eliminated in eastern (Indian) Punjab soon after Independence by the anti-*zamindari* legislation, the absorption of the princely states and the rise to political dominance of the Jutt peasantry. However, in western (Pakistani) Punjab it suffered no such loss (Major, 1991) until 1959, the *jagirs*, the system of *begar* (labour rent or corvee) and the princely states were not abolished. Despite the land reforms of Ayub Khan and Zulfiqar Ali Bhutto, the landed elite managed to retain its former position of control over power structures and continued its colonial legacy of protecting the military interests at the cost of democratic institutions.

COLONIAL ADMINISTRATIVE STRUCTURE AND THE LAND SETTLEMENT POLICY

The Administrative System

Till 1919 Reforms, Punjab was deprived of a constitutional government and maintained as a non-regulation province. There was no Executive Council or High Court. It did not have any effective representation in the Imperial Council. Immediately after its annexation, the province was placed under the authoritarian form of administration by Lord Dalhousie's handpicked British officers, which came to be known as the Punjab School of Administration having a paternalistic attitude towards the province (Major, 1996). Thorburn writes:

> When Lord Dalhousie rounded off the north-west boundary of our Indian domination by annexing the Punjab, he had by right of conquest a clean state to work upon and he then . . . introduced changes in the land revenue system which, however convenient for budgets and disbursements, was fraught with serious consequences for the rural communities of the province. (Thorburn, 1971: 229)

Punjab was a 'clean state' for Dalhousie for 'an imperial experiment, imperially conducted' (Khushwant Singh, 1966). The 'imperial experiment' that was conducted in the Punjab between 1849 and 1856, the period of Lord Dalhousie's governor-generalship, was essentially an experience in authoritarianism and political exclusiveness, which became a trademark of colonial policies in Punjab. Punjab was put under the control of bureaucracy as a non-regulation province, a unique system of government run by a Board of Administration consisting of three members.[32] These three members were given extraordinary powers to settle the land revenue demands, regulate the excise, supervise the police, and impose death penalty for serious crimes. In this form of government, day-to-day business was carried out in accordance with an extremely flexible interpretation of basic administrative instructions contained in printed codes and circulars, instead of observing legislative regulations, as was the case elsewhere in British India.[33] Dalhousie had personal control over the Punjab administration. It was a 'military way of government' carried out by executive orders (Stokes, 1959). Under the Board of Administration were seven commissioners 'to exercise a stricter supervision and control over the deputy commissioners'.[34] The deputy commissioner was the head of the district, the collector of land revenue and also the district magistrate, who exercised his authority and collected revenue largely with the assistance of officially appointed *zaildars* and *lambardars.* All authority, whether executive, revenue or judicial, was concentrated in the hands of the deputy commissioner, who was a 'government' for the majority of his people (Blunt, 1937). Thorburn called him 'a little king within his own domain, subject to loosely defined limitations' (Thorburn, 1971:165). The district was the centerpiece of the British administration, with a village as the lowest rung and the province itself as the highest. In Punjab, the district assumed added significance because it was a non-regulation province.[35] Sir Patrick Fagan described the defining element of the non-regulation provinces like Punjab in the following words: 'The union of all powers, executive, magisterial and judicial, in the hands of the district officer, here termed deputy commissioner in place of magistrate and collector' (Fagan, 1932: 87).

Under the Indian Council Act of 1861, legislative councils were set up in other provinces of India, except Punjab where every effort was made to restraint the establishment of the Punjab Legislative Council. The first Punjab Legislative Council was established in 1897, after a long gap of three decades, and even then it was denied of the powers of interpretation or the privilege of discussing the annual budget given to other legislative councils under the Indian Council Act of 1892. The Punjab Legislative Council was meant to be a loyal body, all its five Indian members were to be nominated by the governor, and the sole criterion of their nomination was 'their loyalty first and then their wealth and rank'.[36] For every legislation, sanctions of the governor-general and the secretary of state were mandatory and the council had very limited powers. The governor was the council's president and empowered to convene the council's meeting. Between 1897 and 1909, it met only 26 times and four British members transacted virtually all its business.[37]

As a result of the Morley–Minto Reforms, the council was expanded. The 1909 Indian Councils Act extended the powers and membership of the council, introducing an element of indirect election into this. However, the majority of its Indian members came from the landed class, who rarely participated in any debate and left the committees' work to the British officials (ibid.). The record of the Legislative Council of the years 1910–12, reveals the lack of participation in the council's proceedings on behalf of the Indian members. Questions were rarely asked about the financial statements presented in the council and no resolution was passed during this period (ibid.). The tradition of non-participation by Indian members in the Legislative Council continued until 1919, when more powers were given to the councils and the membership was enlarged and the concept of a 'restricted franchise' was introduced. Punjab was not allowed to develop democratic institutions like other Indian provinces in the wake of the 1919 Reforms.

COLONIAL LAND SETTLEMENT POLICY

Along with the administrative structure, the 'imperial experiment' of Dalhousie also introduced a 'land settlement policy', which became the major instrument of colonial capitalism in Punjab, through which the colonial rulers transformed the agrarian structure of the Punjabi society. As discussed earlier, in the absence of any 'traces of established leadership', structuring a reliable rural administration was a great challenge for the early British administrators of Punjab. This was achieved by redesigning the Punjabi rural structures and creating a 'tribal social organization' in which 'tribes' became the basis of authority.[38] To safeguard their colonial interests, the British introduced a system 'which generated the worst kind of feudal exploitation of the Muslim peasants' (Sayeed, 1980: 8). Alavi observed that 'the new concept of property in land that was introduced by the colonial regime was premised on dispossession of the peasants' (Alavi, 1980: 371). The traditional self-sufficient village communities were replaced by *zails* headed by *zaildars*, politically and economically dependent on the British authority (Sayeed, 1980).

The 'settlement' was a set of arrangements with a twofold purpose: one, the assessment of land revenue, and two, the framing of a record of rights. Under this 'set of arrangements', Punjab districts were divided into administrative units called *zails* (circles), varying in size from ten to forty villages, with the intention of incorporating tribal organization into their administration by concentrating particular *biradris* in particular *zails* along the lines of tribal kinship residence rules. Each *zail* was headed by a *zaildar* appointed by the British administrators and he was 'the representative of government in his circle'.[39] Very often the existing traditions of lineage or *biradri* leadership were ignored to create a class of tribal leaders who represented British authority in *zails*. In 1872, the Punjab government established the *zaildari* system on a formal basis, laying down the duties and remuneration of the *zaildars*. The *zaildars* and their assistants, the *ala-lambardars* were described by the colonial rulers as the 'channel of communication' or the 'missing link' (Saunders, 1873). Khalid Bin Sayeed observes, 'whatever authority the village community had exercised was disintegrating under the British and gravitating toward the *zaildar*' (Sayeed, 1980: 5). The *zaildari* system replaced the village communities as self-sufficient political and economic units with a political organization based upon extended kinship ties, 'tribes'.[40] *Biradris* in rural Punjab were treated as tribes by British administrators who emphasized the importance of 'tribes' and of extensive supra-village kinship ties, based on tribal affinity, in establishing the secular bonds holding the Punjabi village society together (Gilmartin, 1981).

In arranging the *zails*, the government instructed in 1873 that 'care should be taken to include in one circle as far as possible the people of one tribe—or having some sort of affinity so that discordant elements . . . [could] be reduced to a minimum.[41] Through the delineation of *zails* in accordance with the patterns of tribal settlement, colonial rulers established a framework within which the problem of local leadership could be confronted. *Zaildari* became a vehicle by which the British could en-noble a family, and give it a status of a local leader, a fact noted by many settlement officers.[42] Selection of *zaildars* produced families of considerable local influence, who used their position to claim the leadership of the *zails'* dominant 'tribes'. The granting of *zaildari* represented a means by which a class of rural leaders was created, tied closely to the colonial administration, practising its authority largely in a 'tribal' idiom. In this colonial system of control, the influence of *zaildars* was linked to their relationship with administration and their status dependent on their claims to tribal leadership. The land control and 'tribal' leadership became the central element in establishing the legitimacy of politically effective intermediary elite in rural Punjab. Special *zails* were created to accommodate the 'loyal' families of western Punjab through the granting of *inams* or revenue-free land in each district. The authority of these families was enforced by making their leading men *zaildars.* For instance, in Shahpur district, the British administrators named the 'loyal' Tiwanas and Noons as honorary *zaildars* 'within the estates owned by themselves or their near relations.[43] After 1890, like 'loyal chieftains' of Punjab, the majority of *zaildars* were also converted into a landed class through the wide-scale use of grants of land in the newly created canal colonies.

We have discussed earlier how this class became so vital to the structure of colonial rule in Punjab. Another major instrument in the creation of this class was the Land Alienation Act of 1900, which declared certain 'tribes' as agricultural whose land could not be alienated. The Act, which was based upon the 'tribal idiom', declared that an agriculturist was the one who was born in a particular caste, and not necessarily the actual tiller of the soil.

This definition of an agriculturist along with the land settlement policy defined landed patrons as legitimate links between the colonial state and the populace.[44] Analysing the impact of the Act, Khalid Bin Sayeed writes, 'the Land Alienation Act of 1900 turned out to be a license for land grabbing on the part of big landowners' (Sayeed, 1980: 7). Sir Michael O'Dwyer, the Governor of Punjab, noted: 'It is now regarded by hereditary landowners of all religions and castes as their Magna Carta' (O'Dwyer 1918: 39).

Another aspect of land settlement policy was the colonization of the wasteland and the creation of the canal colonies. To meet the needs of the Indian Army, it was necessary to bring into cultivation the vast arid areas of western Punjab. For this purpose, a huge network of irrigation canals was constructed which converted the wastelands of western *doabs* into 'canal colonies' and introduced commercial agriculture in Punjab. In the pursuit to consolidate their political position, the British used land distribution to fulfil military requirements and maintain an extractive system, which could finance their administrative system. Talbot writes:

> The creation of the Canal Colonies was closely linked with the other major development of the late nineteenth century, the Punjab's emergence as the leading recruitment center of the Indian Army. One of the important incentives for enlistment . . . was the British policy of rewarding ex-servicemen with lucrative grants of land in the colonies. (Talbot, 1988: 40)

Since these colonies were established on the Crown wasteland, the colonial state had the right not only over land distribution, but also over the distribution of water, a new phenomenon in the agricultural life of the province. The canal colonies became the most influential tool in the hands of the colonial administrators to protect the landed and military interests by giving grants of land for military purposes to the class of Muslim landowners committed to the maintenance of the Punjab 'military machine'. Sir Michael O'Dwyer's scheme for grants to the 'landed gentry' which was first developed in 1914 in connection with the distribution of land in the Lower Bari Doab colony, helped to consolidate the landed elite in western Punjab.[45] Under this scheme of 'landed gentry', loyal families of western Punjab including the Tiwanas and Noons of Shahpur and the Tumendars of Dera Ghazi Khan received numerous grants in lieu of their 'loyal' services to the colonial rulers.[46] Commenting on the distribution of land in canal colonies, Imran Ali writes:

> The rural elite of this region was incorporated into the emergent landholding structures of these commercially valuable lands, through several different types of grants. . . . 'Yeoman' grants of 50–150 acres, 'Capitalist' grants of 150–600 acres. . . . Yeoman Horse-breeding and 'stud farm' grants, the 'Landed Gentry' grants. . . . Additional holdings came in the form of 'Reward' grants for administrative, political and military related services. (Ali in Low, 1991: 44)

Ali observes that through these grants, 'concessionary leaseholds' and many other '*ad hoc* arrangements' to acquire land, the new lands were monopolized by the agriculturist elite and the landless rural population was denied a landholding status in the canal colonies (ibid.). 'The strengthening of the upper classes through this extensive process of land acquisition and economic growth provided the cement to bind the Punjabi political structure dominated as it was by the rural interests, with British rule' (ibid.). This process not only resulted in the underdevelopment of the region it also hampered the growth of nationalism. Punjab was the only province of British India whose elite remained loyal till the last days of the empire, against the general wish of the people. The colonial state did not allow the establishment of democratic institutions in Punjab. As a result, no major national political party such as the Indian National Congress and All India Muslim League could play an effective role in Punjab politics. Though the Muslim League emerged as the majority party in the election of 1946, it was not because of the League's organizational skills rather it was the opportunity strategy of a large number of Unionists who joined the Muslim League when they realized that the colonial rule was coming to an end.

NOTES

1. The word 'colonial' refers to the global expansion of trade and conquest by European powers. The process started in the fifteenth century, which conditioned societies in Asia, Africa and Latin America. For elaboration of this term, see Hamza Alavi, 'Colonial and Post-Colonial Societies', in Tom Bottomore (ed.) (1983), *A Dictionary of Marxist Thought*, Cambridge, Massachusets: Harvard University Press.
2. The term 'Third World' is used for countries which were conditioned by European colonialism and got independence after Second World War. The common factor that makes these countries the 'Third World' is poverty and underdevelopment.
3. Irfan Habib, 'The State and the Economy', in Tapan Raychaudhuri and Irfan Habib (eds.). Habib writes that 'a single *pargana*, Fatehpur, contained eight *raiyati* and nine *zamindari* villages'. Land which was not assigned but from which land revenue was extracted by the state was called *khalisa*. Alavi notes that in *zamindari* villages, *zamindars* were responsible for the collection of revenue, while in *raiyati* villages, the peasant community enjoyed a relative degree of autonomy from the *zamindar* or *taluqdar*. In these villages revenue was collected by *muqaddams* (village headmen) (Alavi, 1980). Giving the description of the Mughal ruling class, Habib writes that the Mughal ruling class consisted of about 8,000 *mansabdars* who were responsible for maintaining military contingents available at the service of the state Habib, (1963).
4. According to Habib (1963) there are cases when *zamindari* was purchased at a low sale price. The case is of the English Company's purchase of *zamindari* of villages constituting the nucleus of the future city of Calcutta in 1703 at the sale price of Rs. 1,300 only, when the annual land revenue was Rs. 1,194. Cf. G. Blyn (1964), 'Revenue Administration of Calcutta in the First Half of the 18th Century', *The Indian Economic and Social History Review*, New Delhi, 1(4).
5. Discussing the main features of proprietary rights in Punjab, the *Manual of Land Settlement* produced by early British administrators, states (a) that the right-holder is entitled to the use and occupation of the land during his lifetime; (b) that on his death this title passes to his descendants, subject to customary rules of inheritance, which usually exclude females; (c) that the right-holder is entitled to let the land to tenants on such terms as he thinks fit; (d) that the right-holder can sell or mortgage the land subject to customary and legal restrictions which give to members of the same family or village community a right to interfere under certain circumstances. This

right is based originally on kinship, real or assumed, and not on any claim on the part of the objector to a superior title. Thomason regarded freedom of transfer as a necessary feature of proprietary right. But the Indian idea of property in land is that it is vested in a family, and not in an individual. In many parts of the country the possession of unlimited powers of alienation by the recorded right-holder was entirely opposed to native sentiment, and restrictions on the power of alienation have never been wholly wanting in the Punjab, and have been greatly extended by the Alienation of Land Act, XIII of 1900; (e) that the right-holder is entitled to engage for the payment of the land revenue. This last feature of proprietary right is mainly the creation of our rule' (Moudie, 1908).

6. 'A village community is a body of proprietors who own or formally owned part of the village lands in common, and who are jointly responsible for the payment of revenue. As time goes on the tendency is for the area held in severalty to increase, but it is rare indeed to find a village, which was one of the communal types, in which there is no common property remaining. Joint responsibility has been made a prominent feature of village tenure by the British Government. Under native rule it did not exist when the State realized its dues by division of crop or by appraisement. Even when a cash assessment was made only a few leading members of the community became responsible and they generally occupied the position of revenue farmers in their dealings with the rest of the brotherhood' (Moudie, 1908: 61).
7. Under the Sikh rule, Punjab was divided into several provinces and districts. These provinces were assigned to the 'highest bidders' who squeezed the peasant at their own pleasure. A *nazim* supervised the administration of a given territory, using force to extract the surplus, which was absorbed either by the Sikh army or the *durbar.*
8. Without established institutions, the Sikh state, which was centered around a Sikh oligarchy, had to disintegrate due to internal decay and external pressures.
9. At the beginning of First World War, Punjab had about 11,00,000 men of all ranks in the Army. At the close of the war, no less than half a million had served it. Roughly 3,60,000 Punjabi recruits were enlisted in four years between 1914 and 1918, comprising one half of the total number raised in India. Michael O'Dwyer (1925 and 1923), *India As I Knew It, 1885–1925,* London: Constable & Company Ltd.; *India's Contribution to the Great War,* Calcutta: Government Central Press.
10. L.H. Griffin and C.F. Massy (1940), *Chiefs and Families of Note in the Punjab,* 2 vols., Lahore: Government Printing Punjab, Punjab Archives (henceforth PA). This is the single most comprehensive account of the

social history of the Punjab elite. It contains short stories of the major families of each Punjab district and the Punjab Chiefly States and tells how these families belonging to peasant origin achieved 'elite' status during the Sikhs and British rule. In the Preface to the original edition of the *Punjab Chiefs*, Sir Lepel Griffin stated:

> The intention of the work has been to give a picture of the Punjab aristocracy as it exists at the present day. No mention has accordingly been made of many families, Hindu and Muhammadan, once powerful and wealthy, which fell before the Sikhs. No mention has been made of many old Sikh families whose *jagirs* were seized by Maharaja Ranjit Singh, and whose descendants are now plain husbandmen. A few notices of tribes and families of no present importance have, for special reasons, been given; but general rule, only the histories of those men have been written who possess, at the present time, rank, wealth or local influence'. (Griffin and Massy, 1940: i)

11. *Imperial Legislative Council Debates, 1917–18*, 13 September 1917.
12. During the First World War, heads of some Muslim families of western Punjab showed their loyalty by enlisting recruits for the Indian Army. Most notable amongst them were Sardar Muhammad Nawaz Khan and Sikandar Hayat Khan in Shahpur and Jhelum districts who were assisted by *pirs*, such as Pir Ghulam Abbas of Makhad, Pir Fazal Shah of Jalalpur, Pir Badshah of Bhera, Pir Chan and Pir Sultan Ali Shah of Jehanian Shah. All these *pirs* directed their Muslim followers in the Shahpur and Jhelum districts to enlist. The involvement of *pirs* in raising recruits was very important since the Indian Army was fighting against the Ottoman armies in Mesopotamia. See M.S.D. Butler (1921), *Record of War Services in the Attock District 1914–1919*, Lahore: Government Printing Punjab, PA. *War Services of the Shahpur District*, Lahore: Government Printing Punjab, PA.
13. For the list of tribes listed as 'agriculturist tribes' see S. Gurcharn Singh (1901), *The Punjab Alienation of Land Act XIII of 1900*, Lahore: Government Printing Punjab, PA. For example, in Jhang the agriculturist tribes included Jats, Rajputs, Bilochs, Syeds, Koreshis, Kokaras and Nekokaras. In Multan the list included, Kharals, Awans, Mughals, Pathans, Khokars, Arians, Gujars, Ods and Mahtams in addition to Jats, Rajputs, Bilochs, Syeds and Koreshis.
14. For instance, Pir Sultan Ali Shah, head of the Syeds of Jehanian Shah in Jhang, received five squares as a 'landed gentry' grant, a First Class *Khillat* in 1917 and a Sword of Honour in 1919. Syed Ghulam Muhamad Shah

was given a seat on the Bench of Honorary Magistrate at Sahiwal for his 'hard work to induce the people of their *zail* to enlist, and met with a very fair measure of success.' *War Services of the Shahpur District*, p. 37.

15. For details of these rewards see Chopra, *Chiefs and Families*, vol. 1, pp. 195–6, 220, 250–1, 356, 363–4, 416–17, 468–9, 492, 500; vol. 2, pp. 40, 91, 108–9, 118, 203–5, 224, 234, 238, 248, 276, 290–1, 294, 315, 324, 32, 331, 333–4, 339, 345, 359–62, 369, 376–7, 381, 385, and *Titled Gentlemen and Chiefs other than Ruling Chiefs*, Lahore: Government Civil Secretariat Press, 1876, pp.100–1, 112–13, PA.
16. *Revenue Department Proceedings*, 3 July 1858, no. 44, PA. In this letter dated 8 June 1858, G.F. Edmonstone, Secretary to the Government of India with the Governor-General, wrote to the Chief Commissioner of Punjab:

    As it is probable that, in rewarding the good services of those who may have distinguished themselves in aiding the British Government during the late disturbances, grants of land, either entirely or in part rent-free, may be made to such persons in perpetuity or for a term of years, I am directed to intimate that, in the opinion of the Right Hon' ble the Governor General, it is very desirable that these grants should be conferred by a written deed, and that an express condition should be inserted, making the continuance of the grant dependant on the good conduct of the grantee and his successors, the holding being liable to resumption in case of any violation of this condition.

17. *General Report on the Administration of the Punjab Territories for the year 1865–6*, para 24, Lahore: Government Printing Punjab, PA, 1866.
18. For rules for guidance of honorary magistrates, see *Judicial Department Proceedings*, 11 January 1862, nos. 12–12.5, PA.
19. *Judicial Department Proceedings*, 23 November 1867, no. 41, PA.
20. As a Court of Wards, the deputy commissioner was empowered to take charge of the management of the estates of females, minors, idiots, lunatics and inheritors considered otherwise to be unfit for direct management. In the case of minors, the jurisdiction of the Court of Wards extended to the education of wards. The powers of deputy commissioners and commissioners conferred on them under Section 4(3) of the Punjab Court of Ward Act, II of 1903. See Rules of General Procedure under Section 54 of the Punjab Court of Wards Act, II of 1903, Court of Wards, Account Code, Lahore: Government Printing Punjab, PA, 1935.
21. *Report of the Administration of estates under the Charge of the Court of Wards for the Year Ending 30 September 1895*, Lahore: Government

Printing Punjab, PA, 1896, and *Punjab Departments Annual Reports*, L5 VI (3), IOR. For instance, the Tiwana family of Kalra estate in Shahpur proved to be the most loyal family of the Raj. Khizer Hayat Tiwana as a minor heir was put under the supervision of the colonial state and his estate Kalra was developed as one of the largest and most influential estate under the control of the Courts of Wards Administration.

22. 'A year ago I told you with pride that since the War began the Punjab had furnished 124,000 men. That was soared over 2 & ½ years. What have we done within the last year? We have furnished 127,095 combatants or more than in the previous 2 & ½ years. All the rest of India including the Native States with more than 12 times our population have raised in the last 2 & ½ years 137,000 or slightly more than one single province. Since the war began, we have raised 250,000 men to fight the battles of Empire, besides some 60,000 or 70,000 men serving as non-combatants. . . . It is a truly remarkable one for a single Indian province.' Speech of Sir Michael O'Dwyer, Lieutenant Governor of the Punjab, on 26 April 1918 in the Punjab Legislative Council, *War Speeches of Sir Michael O' Dwyer, Lieutenant Governor of the Punjab*, Lahore: Government Printing Punjab, PA, 1918.
23. In Rawalpindi and Jhelum districts, over 30 per cent of the male population went to the war, in Attock district the figure was 16 per cent in, Gujrat 13 per cent and in Shahpur 10 per cent. See 'Speech by Sir Michael O' Dwyer in the Punjab Legislative Council, 26 April 1918', *War Speeches of Sir Michael O' Dwyer Lieutenant Governor of the Punjab.*
24. As a reward, generous land grants were given to the servicemen in 'canal colonies'. At the end of the First World War, over 4,20,000 acres of colony land was distributed to just over 6,000 commissioned and non-commissioned army officers. See B. Josh (1979), *Communist Movement in the Punjab 1921–47*, New Delhi: Peoples Publishing House. Similarly Muslim landowners and *pirs* who were instrumental in raising recruits were also given land grants in the canal colonies. These land grants became the major tool through which the landed class was created in the Punjab. See Imran Ali, *The Punjab Under Imperialism, 1885–1947*; and *War Speeches of Sir Michael O' Dwyer, Lieutenant Governor of the Punjab.*
25. 'The Tumendars have recently received liberal grants of land from the *Sarkar* in the Lower Bari Doab Colony in recognition of their position and past services. These grants are conditioned on active loyalty, and I am confident that they will prove this by redoubling their efforts to raise recruits among their tribes.' Speech of Sir Michael O'Dwyer, Lieutenant Governor of the Punjab, delivered at the Special Durbar held at Dera

Ghazi Khan on 18 February 1918, *War Speeches of Sir Michael O' Dwyer, Lieutenant Governor of the Punjab*, p. 110.

26. Franchise proposals of the Punjab Government, L/ Parl./409B, pp. 147–205, IOR.
27. Ibid. In the urban areas the number of seats was restricted to a bare minimum and very high voting qualifications were proposed. These qualifications restricted the vote to 3 per cent of the urban population. *Franchise Proposals of the Punjab Government.*
28. Ibid.
29. GI Reforms Office, Deposit, April 1920, no. 5: Proceedings of the Governor's Conference on the Reforms, 23 January 1920, L/ Parl./ 409B, IOR.
30. GI Reforms Office, General, Deposit, Dec. 1919, no. 12: Memorandum by Umar Hayat Khan Tiwana, July 1919, L/ Parl./ 409B, IOR. His son Khizer Hayat Tiwana became the second premier of the autonomous Punjab in 1942. F. Popham Young, Commissioner of Rawalpindi, writes in his comments: 'the Tiwana Maliks feel, what is indeed the truth, that Government has made them what they are, and they have responded by providing themselves to be loyal supporters of Government not with mere passivity but with real activity.' Comments of F. Popham, *Register of Landed Gentry Grants*, Rawalpindi Division, 15 (Punjab Board of Revenue, file 301/1176KW).
31. Talbot (1988). According to the 1921 census, the communal ratio of population was Muslims 55.33 per cent; Hindus 33.58 per cent and Sikhs 11.00 per cent. The ratio of Muslim voters was highest in rural areas having a record of recruiting servicemen, i.e. voters in Jhelum and Rawalpindi composed 14 per cent of the Muslim electorate in the Punjab. See *Reports of Indian Statutory Commission*, vol. 2 , London: His Majesty Stationery Office, 1930 [Survey CMD 3568], IOR.
32. 49 British civil and military officers of the Company were chosen to serve under the Board headed by General Henry Lawrence.
33. Civil and criminal justice and land revenue administration was based upon codes, circulars and rules issued by Dalhousie to the Board of Administration. See Dalhousie's instructions to the Board of Administration, *Selected Circular Orders of the Board of Administration in the General and Political Departments*, Lahore: Government Printing, Punjab, 1871, PA.
34. *Selected Circular Orders of the Board of Administration in the General and Political Departments*, Lahore: Government Printing Punjab, 1871, PA.
35. The distinction between regulatory and non-regulatory systems was simple. In the regulatory system, the province was to be regulated by an Act of

Parliament or Regulations of the Government of India. In 1793, Lord Cornwallis issued a revised code of 48 regulations for the presidency of Bengal. This body of legislation, subsequently amended, was known as the Bengal Regulations. In the regulation provinces the head of a district was a judge and magistrate, with powers to adjudicate civil disputes and minor criminal cases. More serious cases were handled by the provincial courts of appeal and circuit. The chief civil and criminal courts at Calcutta controlled these courts. In the non-regulation provinces, the procedures were simple, providing discretionary powers to the official in the conduct of administration.

36. *Punjab Administrative Reports 1922–3*, Lahore: Government Printing Punjab, p. 17, PA.
37. See *Punjab Legislative Council Debates 1897–1936*, V9 3419–V9 3455, IOR.
38. For a detailed discussion on the British administration and the tribes, see Gilmartin (1988).
39. *Proceedings of the Lieutenant Governor*, 29 February 1872, Punjab Board of Revenue, file 61/142.
40. See C.L. Tupper (1880), *Punjab's Customary Law: Statements of Customary Law in Different Districts*, vol. II, Simla: Government Central Branch Press, PA. David Gilmartin (1981), 'Kinship, Women and Politics in Twentieth Century Punjab', in Gail Minault (ed.), *The Extended Family: Women and Political Participation in India and Pakistan*, Columbia: Columbia University Press.
41. Government letter to settlement commissioner, Delhi division, 30 June 1873, Punjab Board of Revenue, file 61/142.
42. P. J. Fagan, Jr. (secretary to financial commissioner to revenue secretary) 25 October 1890, Punjab Board of Revenue, file 61/134.
43. *Shahpur District Gazetteer, 1917*, p. 243, PA.
44. 'Agricultural Tribes' were gazetted by name in each district. The status of these tribes was to be determined by their agricultural character and by their political importance to the British. N.G. Barrier (1966), *The Punjab Alienation of Land Bill of 1900*, Durham: Duke University Press, p. 112.
45. See Note by Sir Michael O'Dwyer, Lieutenant Governor, on the Lower Bari Doab Colonization Scheme, 8 August 1913, Punjab Board of Revenue, file 301/1176.
46. See *Register of Landed Gentry Grants* (for grants to the Tiwanas), Rawalpindi division (for grants to the Dera Ghazi Khan Baloch); Multan division, PA.

REFERENCES

Adelmann, I. and C.T. Morris (1973), *Economic Growth and Social Equality in Evolving Countries,* Stanford: Stanford University Press.

Alavi, H. (1983), 'Class and State in Pakistan', in Hassan Gardezi and Jamil Rashid (eds.), *Pakistan: The Unstable State,* Lahore: Vanguard Books.

——— (1980), 'India: Transition from Feudalism to Colonial Capitalism', *Journal of Contemporary Asia,* 10: 4.

——— (1973), 'The State in Post-Colonial Societies: Pakistan and Bangladesh, in Kathleen Gough and Hari P. Sharma (eds.).

Ali, Imran (1988), *The Punjab Under Imperialism, 1885–1947,* Princeton: Princeton University Press.

Almond, Gabriel and J. Coleman (eds.) (1960), *The Politics of the Developing Areas,* Princeton: Princeton University Press.

Amin, Samir (1976), *Unequal Development: An Essay on the Social Formation of Peripheral Capitalism,* New York: Monthly Review Press.

——— (1973), *Neo-colonialism in West Africa,* New York: Monthly Review Press.

Baden-Powell, B.H. (1892), *Land Systems of British India,* Oxford: Clarendon Press.

Bauer, Peter T. and Basil S. Yamey (1978), 'The Third World and the West: An Economic Perspective', in W. Scott Thomson (ed.).

Binder, Leonard (1961), *Religion and Politics in Pakistan,* Berkeley: California University Press.

Blunt, Edward (1937), *The I.C.S.: The Indian Civil Service,* London: Faber & Faber.

Chandra, Bipan (1980), 'Colonialism, Stages of Colonialism and the Colonial State', *Journal of Contemporary Asia,* 10: 272.

Cohen, Bernard S. (1971), *The Social Organization of a Civilization,* New Jersey: Englewood Cliffs.

Cohen, Stephen P. (1984), *The Indian Army: Its Contribution to the Development of a Nation,* Berkeley: University of California Press.

——— (1983), *The Pakistan Army,* Berkeley: University of California Press.

——— (September 1980), *Security Decision-Making in Pakistan* (Report prepared for the office of External Research Department of State, Contract # 1722-020167, University of Illinois, Urbana, Unclassified monograph, Copy # 45).

Escobar, A. (1985), 'Discourse and Power in Development: Michel Foucault and the Relevance of His Work to the Third World', *Alternatives,* 10: 227–400.

Fagan, Patrick (1932), 'District Administration in the United Provinces, Central Provinces and the Punjab, 1818–57', in H.H. Dodwell (ed.), *The Cambridge History of India, Volume VI: The Indian Empire 1858–1918*, Cambridge: Cambridge University Press.

Fernando H. Cardoso and Enzo Falettu (1978), *Dependency and Development in Latin America*, Berkeley and Los Angeles: University of California Press.

Frank, Andre G. (1981), *Crisis in the Third World*, New York: Homes and Meier.

——— (1979), *Dependent Accumulation and Underdevelopment*, New York: Monthly Review Press.

——— (1972), 'The Development of Underdevelopment', in James. D. Cockroft et al. (eds.), *Dependence and Underdevelopment: Latin America's Political Economy*, New York: Anchor Books.

——— (1968), *Capitalism and Underdevelopment in Latin America*, Harmondsworth: Penguin.

——— (1966), 'The Development of Underdevelopment', *Monthly Review*, 18: 17–19.

Franke, R.W. and B.H. Chasin (1980), *Seeds of Famine*, New York: Seabury Press.

Gallagher, J., J. Gordon and A. Seal (1973), *Locality, Province and Nation: Essays on Indian Politics 1879–1940*, Cambridge: Cambridge University Press.

Gardezi, Hassan and Jamil Rashid (eds.) (1983), *Pakistan: The Unstable State*, Lahore: Vanguard Books.

Gardezi, H. (1983), 'Feudal and Capitalist Relations in Pakistan', in Hassan Gardezi and Jamil Rashid (eds.).

——— (1973), ' Neo-Colonial Alliances and the Crisis of Pakistan', in Kathleen Gough and Hari P. Sharma (eds.).

Gilmartin, David (1988), *Empire and Islam: Punjab and the Making of Pakistan*, London: I.B. Tauris.

Gough, Kathleen and Hari P. Sharma (1973) (eds.), *Imperialism and Revolution in South Asia*, New York: Monthly Review Press.

Griffin, K. (1978), *International Inequality and National Poverty*, London: Macmillan.

Habib, Irfan (1984), 'The State and the Economy—Mughal India', in Tapan Raychaudhuri and Irfan Habib (eds.).

——— (1984), 'Agrarian Relations and Land Revenue', in Tapan Raychaudhuri and Irfan Habib (eds.).

——— (1969), 'Potentialities of Capitalistic Development in the Economy of Mughal India', *The Journal of Economic History*, XXIX.

———(1963), *The Agrarian System of Mughal India (1556–1707)*, Bombay: Asia Publishing House.

Haq, Mahbub ul (1973), *The London Times*, 22 March quoted in Mahbub ul Haq, *The Poverty Curtain: Choices for the Third World*, New York: Columbia University Press.

Hassan, Nurul (1969), 'Zamindars under the Mughals', in Richard E. Frykenberg (ed.), *Land Control and Social Structure in Indian History*, Madison: University of Wisconsin.

Huntington, Samuel P. (1968), *Political Orders in Changing Societies*, New Haven: Yale University Press.

La Palombara, Joseph (ed.) (1963), *Bureaucracy and Political Development*, Princeton: Princeton University Press.

Low, D. A. (1991) (ed.), *The Political Inheritance of Pakistan*, London: Macmillan.

MacMunn, George F. (1979), *The Martial Races of India* [1930], Delhi: Mittal Publications.

Major, Andrew (1996), *Return to Empire: Punjab Under the Sikhs and British in The Mid-Nineteenth Century*, Karachi: Oxford University Press.

———(1991), 'The Punjab Chieftains and the Transition from Sikh to British Rule', in D.A. Low (ed.).

Moudie, J.M. (1908), *The Punjab Administrative Manual*, Lahore: Government of India, Punjab Archives.

Murdoch, William W. (1980), *The Poverty of Nations*, Baltimore: Johns Hopkins University Press.

Naseem, S.M. (1980), *Underdevelopment, Poverty and Inequality in Pakistan*, Lahore: Vanguard Books.

Ouselev, G. and W.G. Davies (1886), 'Report on the Revised Settlement of the Shahpoor District in the Rawalpindi Division', Lahore: Government Printing Punjab.

Page, David (1987), *Prelude to Partition: The Indian Muslims and the Imperial System of Control 1920–1932*, Karachi: Oxford University Press.

Pasha, Mustafa Kamal (1998), *Colonial Political Economy: Recruitment and Underdevelopment in the Punjab*, Karachi: Oxford University Press.

Raychaudhuri, Tapan and Irfan Habib (eds.) (1984), *The Cambridge Economic History of India, Volume 1, c. 1200-c. 1750*, Hyderabad: Orient Longman in association with Cambridge University Press.

Rehnema, Majid (1986), 'Under the Banner of Development', *Development*, 3: 47–67.

Santos, Theotonio dos (1984), 'The Structure of Dependence', in Michelle A. Seligson (ed.)

Sayeed, Khalid Bin (1980), *Government and Politics: The Nature and Direction of Change*, New York: Praeger Press.

Saunders, L.S. (1873), *Report on the Revised Land Revenue Settlement of the Lahore District in the Lahore Division of the Punjab, 1865–69*, Lahore: Government Printing Punjab.

Seligson, Michelle A. (1984) (ed.), *The Gap Between Rich and Poor: Contending Perspectives on the Political Economy of Development*, Boulder and London: Westview Press.

Singh, K.S. (1978), 'Colonial Transformation of Tribal Society in Middle India', *Economic and Political Weekly*, 29 July.

Singh, Khushwant (1966), *History of the Sikhs*, vol. 2, *1839–1964*, Princeton: Princeton University Press.

Smith, Anthony (1978), 'The Case of Dependency Theory', in W. Scott Thomson (ed.), *The Third World: Premises of U.S. Policy*, San Francisco: Institute of Contemporary Studies.

Sovani, N.V. (1954), 'British Impact on India After 1850–7', *Journal of World History*, April.

Spear, Percival (1965), *A History of India*, vol. 2, Harmondsworth: Penguin Books.

Stokes, Eric (1959), *The English Utilitarians and India*, London: Oxford University Press.

Talbot, Ian (1990), *Pakistan: A Modern History*, Lahore: Vanguard Books.

——— (1988), *Punjab and the Raj: 1847–1947*, New Delhi: Manohar.

Thorburn, S.S. (1971), *The Punjab in Peace and War*, New York: AMS Press.

Valenauela, S. and A. Valenzuela (1984), 'Modernization and Democracy: Alternative Perspectives in the Study of Latin American Underdevelopment', in Michelle A. Seligson (ed.).

——— (1974), *The Modern World System: Capitalist Agriculture and the Origins of the European World Economy*, New York and London: Academic Press.

Wallerstein, I. (1974), *The Modern World System: Capitalist Agriculture and the Origins of the European World Economy*, New York and London: Academic Press.

Wallerstein, I. (1984), 'The Present State of the Debate on World Inequality', in Michelle A. Seligson (ed.).

CHAPTER 3

Globalization and Agriculture: Some Propositions

Aijaz Ahmad

When I was invited to deliver the inaugural address for a conference on the Punjab peasantry, I was greatly puzzled. I have no expert knowledge either on Punjab or on the peasantry in general, nor much of non-expert familiarity; indeed, I have published nothing on the subject and have no competence to do so. I was persuaded on two grounds: that the invitation comes out of deep affection (*Patiala mein aapke chahnewale bahut hain* [you have many admirers in Patiala]), and that I was expected to speak not on the peasantry but on globalization which is the framing context for evolving, crisis-ridden agrarian relations not only in Punjab or India but across a vast swath of countries in Asia and Africa. I shall therefore speak mostly of what gets termed as 'globalization' in contemporary academic writings as well as of 'neoliberalism' which serves as its twin. *Inter alia,* I will also offer brief reflections on some broad implications of these phenomenon for the subject at hand.

My starting point here is that the term 'globalization' is inaccurate, more inaccurate than Ernest Mandel's famous term 'Late Capitalism'. What we should be talking about is a new kind of imperialism that began after the Second World War and lodged itself in place during the past decade or so, since the dissolution of the socialist project in the Soviet Union and elsewhere. Furthermore, this is not *simply* a new phase of imperialism but a *radically* new phase which could not have come about until after a number of preconditions had been met. What were the most important of those preconditions? I shall answer that question in the form of twelve interlinked propositions, and what I want to argue is that it is the sum total of all these elements—not some, not this or that—which came to define the radical novelty of this phase.

The first of those preconditions, I believe, was the dissolution of the colonial empires themselves. The beginning of globalization can thus be dated back to the years following the Second World War. Now, Marx was the first to point out that the drive toward an integrated world market has been inherent in the logic of capitalism from the beginning and that colonialism played the main role in creating such a market. However, between the end of the fifteenth century, when it all began, until the end of the eighteenth century, the process of full territorial conquest was mostly centred on the Americas. It was only in the late eighteenth and nineteenth centuries that the interiors of Asia and Africa were extensively occupied, eventually dividing the world into a set of core industrialized countries of the advanced West and a vast hinterland of non-industrialized colonies and dependencies, many of them formally independent. In the days of its final glory, colonialism had created something resembling a world economy but it was a system, really, of interlocking economies in which different colonial powers controlled different segments. Decolonization was now necessary for the further development of capitalism as a wholly integrated global economy in this new phase, and the US could not have emerged as an unchallengeable hegemonic power until after its rivals had lost their respective colonial empires.

This then leads to the second precondition, namely that there had to be a pre-eminent power to supervise the transition from a number of large and small empires to a united global empire. The quarter century after the Second World War which has gone down in history as the golden era of US hegemony corresponds to the organization of that transition. Much has been said recently about the decline of that hegemony and it is certainly true that other centres, notably western Europe and Japan, also emerged to challenge some aspects of that unique power. It is also true that the US State and capital are beset by numerous crises: the quagmires in Iraq and Afghanistan, mounting public and private debt, declining dollar, reliance on Japanese and Chinese banks to cover budgetary deficits, and so on. It is worth saying, however, that the United States continues to be the most powerful capitalist economy; that it dictates terms in such transnational institutions as the World Bank and WTO; that the US commands more military power than all of western Europe, Russia and Japan combined; that NATO interventions are really American interventions; and that the US has by far the pre-eminent—and I might add, devastating—role in training the techno-scientific managerial elite of the Third World.

Third, as I just said, the era of classical colonialism divided the world into a core of industrialized countries and a vast hinterland of non-industrialized zones. For capitalism to really take off as a globalized system of industry and finance, an altogether new kind of division of the world was necessary, between the advanced and the backward capitalist zones. The dissolution of the colonial empires made possible the national bourgeois project of substantial degree of industrialization in virtually all the zones of the Third World outside Sub-Saharan Africa, and thus vastly altered the very scope of capitalism as such. Not the least aspect of the globalization of capitalism was the subsumption of virtually all the agriculture of the world into the laws of capitalist value.

Fourth, the existence of the Soviet bloc and the East Asian socialist and quasi-socialist states obstructed 'globalization' in three ways. They constituted roughly a third of the world in demographic terms, and that one-third was simply not available for capitalist globalization in the real sense, whatever other relations with the world capitalist market those states might have maintained (witness, for example, the history and mechanics of debt formation among the weaker of the Council for Mutual Economic Assistance (COMECON) countries such as Poland well before the dissolution of the system). Elsewhere, the support they offered was often decisive for wars of liberation during that period, from Vietnam to Angola and Guinea-Bissau, while they also served as alternate sources of technology, training, finance and military supplies for many of the non-aligned countries as well, such as Egypt or Syria or India. For us in India, actually, the kind of development we witnessed in key areas of the public sector, from steel to petroleum to petrochemicals, during the first quarter century after Independence would have been inconceivable without that alternate source. Equally crucially, that sense of an alternate pole in international relations helped sustain a certain left-of-centre ideological position for those post-colonial national-bourgeois states which were so inclined.

The fifth precondition for the full emergence of globalization was the defeat and demise of the radical-nationalist and socialist projects across the Tricontinent (Asia, Africa and South America). For roughly a quarter century after the Second World War, economic nationalism in substantial parts of the Third World had presented something of a barrier to truly global triumph of neo-liberalism through a variety of strategies, such as nationalization of foreign monopolies, protectionism, use of the state sector for growth of domestic industry, relations with the COMECON countries as alternate source for economic aid and technology, and so on. For most of the Third World, that phase had ended by the mid-1970s, in a variety of ways. In countries such as Indonesia (1965) and Chile (1973), it took the form of military coups and bloodbath of millions; most Latin American countries—Brazil and Bolivia, Argentina and Guatemala, to name just four—passed through this experience. A number of progressive African countries—the Congo, Angola, Mozambique—were tamed through a combination of coups and invasions. For Nasser's Egypt or Baathist Syria, Israeli invasion and victory of 1967 proved decisive. In India, the process was bloodless, but the radical phase of the Nehruvian state was in any case over by the time Indira Gandhi declared the Emergency. The final exhaustion of the nationalism of the national bourgeoisie in the Third World coincided then with the dissolution of the Soviet system in eastern Europe and the USSR as well as the promulgation of the policies of capitalist restoration in China. India itself did not turn fully toward neoliberal policies until after the breakup of the Soviet Union. Elsewhere, neither the Sandanista in Nicaragua nor the left within the ANC could sustain its original socialist projects in the face of such debacles. Those who think of globalization as a process that has evolved peacefully out of the purely economic logic of capitalism would do well to recall the histories of the immense violence that was required to suppress Third World nationalism on the global scale.

Sixth, for the neo-liberal regimes of the core imperialist countries to take hold of the globe fully they had to take hold, first of all, of their heartlands, for which the dismantling of the post-war Keynesian compact of high wage, high unemployment regimes, which had passed under the rubric of European social democracy, had to be dismantled. That kind of regime had been generalized in western Europe under a peculiar set of circumstances. The immense destruction of capital assets during the Second World War necessitated reconstruction which gave Europe its longest wave of prosperity in the twentieth century requiring such massive employment policies that not only the European workers but also significant numbers of workers from former colonies could be absorbed into the labour market, while prosperity itself made high levels of money wages as well as social wage possible. Within this structure, of course, high wage for the white working classes was itself tied to generalized low wage for the non-white immigrant labour. Politically, this high employment, high wage regime of accumulation was found necessary to fend off the communist threat, a matter that was not to be settled in south-western Europe until after the electoral defeat of the communists in Italy and rolling back of the revolution in Portugal during the 1970s, not to speak of the collapse of the common programme in France. It was, in other words, the spectre of communism that forced the European capitalists to provisionally suspend the normal capitalist pursuit of minimization of wages and maximization of surplus value, and capitalist stability despite the suspension was underwritten by very high growth rates.

The irony of European developments was that the communist threat disappeared roughly at the same time that the capitalist system was entering into a period of prolonged stagnation, so that neo-liberal assault could now begin. This historical coincidence made it much easier to reverse those earlier tendencies, putting in place the labour-saving information technologies that were in any case favourable to the petty bourgeoisie and mobilizing this petty bourgeoisie in favour of cutting down employment rates as well as money wages for the working class while also cutting the taxes that were necessary to keep up the levels of the social wage.

Seventh, there is the powerful element that gets almost exclusive emphasis in most narratives of globalization, namely the new technologies that were required to integrate the world financial markets globally and to make even productive capital itself relatively more mobile. This has involved, among other things, the creation of an 'Information Super-highway', beyond the territorial boundaries of nations and consisting of millions of pathways. The 'virtual' character of much of this flow has often been exaggerated; it would be more accurate to say that the kind of global production and exchange which these flows facilitate also greatly intensify and extend processes of commodification as well as globalized labour regimes of a historically new kind. In this context, then, neither a world economy nor a world culture, neither global sport nor global entertainment, was possible so long as goods, including financial goods, were exchanged only between individual nations. For globalization of these phenomena, technologies are needed which can beam the same signal into households across the globe and which bring down the travel time for transmission of financial decisions effectively to zero. One might even suggest that the immense technological innovations of the post-war period have been as necessary for the launching of this new, globalized imperialism as the technologies of the second Industrial Revolution had been necessary for stabilization of colonial conquest. Conversely, globalization of these technologies for production and profit would have been inconceivable without the prior existence of proletariats across the globe on a scale unimaginable in the colonial period.

Eighth, this whole edifice is upheld by a complex system of laws and regulations which has two overlapping aspects. There is first of all the regulatory regimes of the IMF, the World Bank, the Industrial Fasteners Institute (IFI), GATT, WTO, and so on which are together fast emerging as a new world government for imposing uniform policies, obligations, and conditionalities around the world, especially the imperialized world. These institutions have been central in perfecting this system, and they of course have their own very complex legal frameworks and regulatory regimes that individual nation states have to abide by. But an equally crucial aspect of this globalization of law and sovereignty is that national governments are being constantly pressed to alter their own laws so as to make them more compatible with, often mere facsimiles of, American law.

Ninth, this new globalized imperialism also required new types of military technologies, the famous 'automated battlefields', for example, which could deliver imperial power effectively and swiftly against various little enemies that were perceived to be proliferating all over the world, as we have recently seen in Iraq and the former Yugoslavia, both of which have served as test cases both for military technologies and for moral claims of the imperial powers. The deployment of these technologies is supervised and even executed primarily by the United States, though often in the name of a wider agency. If Iraq was invaded in the name of the UN, the war over Kosovo was conducted under the aegis of NATO. That same 'automated battlefield' is now in operation in Afghanistan, is threatening to be made operational in dozens of other countries, put into place by the US and UK but with the assistance of numerous countries from Azerbaijan to Australia, and in the name of a fully globalized coalition. Indeed, the unity of this coalition closely resembles the unity of today's globalized market.

Tenth, just as supra-national institutions such as the World Bank and WTO are used to impose new imperial regimes of law and regulation, various institutions associated with the UN, most notably the Security Council itself, are increasingly in use for the moral legitimation of imperial aggression. The use of the United Nations to legitimize American military designs is as old as the Korean War of the 1950s. Then, in the period of revolutionary upsurge of the next twenty years, this unholy alliance receded. For a transitory moment in the mid-1970s, just about the time of the liberation of Vietnam, the UN had even tried to catch up with the revolutionary temper of the times by enacting a Charter of Economic Rights and Duties of States which proclaimed that member-nations had the right not only to 'regulate and exercise authority over foreign investment' and 'the activities of multinational corporations' but even to 'nationalize, expropriate or transfer ownership of foreign property'. Thus, in the context of a different balance of global force, obtained by the revolutionary uprisings and the radical nationalisms of the 1960s, the UN had itself endorsed all the fundamental aspirations of Third World economic nationalism. Those days are no more. The Security Council again endorses, in the name of the 'international community', whatever the US wants, while the General Assembly of the UN has ceased to be a meaningful forum.

Eleventh, in addition to the legal legitimation of American aggression by the Security Council, a complex network was required for moral pressure, ideological legitimization and cultural acceptance among the intelligentsias of the world. This was obtained in great many forms, ranging from all kinds of NGOs to high-minded post-modernism to the 'End of History' ideology and the 'Clash of Civilization' thesis. This moral economy undoubtedly has its own codes and flexibilities. Democracy and human rights are absolute values, except when it comes to great many places like Saudi Arabia and other US allies. The drug-dealing and murderous fanatics of Afghanistan get portrayed as moral agents, even *mujahideen* in the great Islamic cause, so long as they serve as foot-soldiers in America's global anti-communist crusade, but these same *jihadis* come to personify the worst evils of what has come to be called 'Islamic fundamentalism' and now 'terrorism with a global reach' when it suits the new American priorities.

Twelfth, the twin cities of Washington DC and New York serve as the combined capital city of this empire because they are the headquarters not only of the US government, which is the decisive fact, but also of most key institutions of this imperial management: the Wall Street, IMF, the World Bank, WTO, the UN, and so on. Meanwhile, the idea of citizenship is being revised. The collapse of Keynesianism in the core countries and of reformist nationalism in the Third World means that social and economic entitlements are now tied less to citizenship or to employment or to basic survival needs, even in theory, and far more to the market as the final arbiter of the social good. Extreme class polarization produced by new imperialism on a global scale means that the upper-most stratum commands something like a global citizenship, with few barriers to travel residence common consumption patterns and educational facilities—whether the individual member of the stratum lives in Bombay, Sao Paulo, Tokyo, Frankfurt, New York, or Shanghai.

On the other hand, the denial of basic entitlements now increasingly envelop considerable numbers of those wretched of the earth who live in those same cities, but in their ghettos, barrios, *chawls*, and in the vast rural hinterlands. There are substantial islands of the Third World in virtually all the major cities of imperial dominance such as New York, London or Paris, but equally widespread are islands of the First World in cities of the Third World. A globalized, extra-territorial class system, so to speak.

The precondition is the massive reorganization of culture. The central fact here is that we are witnessing the emergence of a worldwide capitalist civilization in which national, regional and local cultures are being reorganized as variants of that singular civilization. Civilizational homogeneity exists at the deepest level, the level of commodification. But at the level of second-order reality—that is to say, in the phenomenal form of the commodity—all kinds of differences are maintained, encouraged, even manufactured. This is because without that diversification, the illusion of freedom and choice cannot be maintained, which is the very essence of the market. That is why ethnicity is such a major component in postmodern thought as well as contemporary fashion design, even in the national markets and cultural complexes of the Third World that have been rearranged in the image of the metropolitan ones. This cultural globalization takes great many forms. That the upper strata of the intelligentsia is trained in a small number of metropolitan universities with the US playing the central role has produced a remarkable degree of ideological homogeneity among them, and in societies under their supervision. As a result strikingly similar patterns of cultural desire extend from Taiwan to Brazil. Meanwhile, the technologies of cultural globalization have a remarkably invasive reach. If the colonial classroom took in a miniscule minority of the population and was restricted to very little public space, the reach of TV channels and mass advertising, hence the cultural patterns they advocate and the consumer goods they sell, goes deep into the household and the rural hinterland. If the colonial classroom sought to modulate the thought process of mainly the privileged strata in the colony, the mediatic culture of globalization does all that but also seeks to colonize the unconscious of the masses of people.

It is in this much broader context that the ideology of globalization has produced much nonsense on the question of the nation-state and its supposed demise in the face of capitalist universalization. In reality, the nation-state form seems to be alive and proliferating. The US has a military establishment that is national and most jingoistically *nationalist.* Germany, the most powerful member of the European Union, has recently achieved its expanded national consolidation by virtually colonizing the eastern zones. The spectacular growth of the so-called 'East Asian tigers' was unthinkable without the strong interventionist nation-states. Policies of neo-liberal privatizations are carried out across the world by powerful governments of individual nation-states.

The Indian government that is at present (2005) headed by Manmohan Singh is, in this sense, no less interventionist than was the Nehruvian State. What has declined is not the power of the State but commitment to at least some measure of redistributive justice. Indeed, as reformist nationalisms as well as the working classes are put so much on the defensive, it is the *States* that have returned so aggressively to enforce policies on behalf of the capitalist classes, giving up even the pretence of being *above* classes and representing the whole nation, *the people.* Capital wants no national barriers for itself but labour regimes are always national, even for immigrants as they move from one national regime to another; you only have to ask Indian workers in Bombay, then in the Gulf kingdoms, and then London or New York. What globalization does want is a nation-state that is *weak* in relation to labour but *strong* on behalf of capital.

Thus, far from being in contradiction with globalization, or declining in the face of globalization, the nation-state form remains the chief instrument of delivering services in actually existing capitalism. What has happened, rather, is that it has been radically detached from its reformist claims of the post-war period in Europe and the first few years after decolonization in Asia and Africa. Global neoliberalism was impossible without this re-structuring of the nation-state, under US hegemony. What has been the effect of this re-structuring?

At the broadest level, Prabhat Patnaik has summarized 'three important consequences: first, a weakening of the working class *vis-à-vis* the capitalists (so much so that many American economists see Reaganomics essentially as a strategy for raising the rate of surplus value); second, a considerable centralization of capital, i.e. the driving out or absorption of small capitalists by the large; and third, a strengthening of financial interests relative to manufacturing interests within the capitalists themselves. In other words, redistribution of incomes *upward*, from the working class to the capitalists, from the smaller to the larger capitalists, and from manufacturing to finance capital; indeed, finance capital seems to benefit from all aspects of these redistributive schemes. Quite aside from the issue of its own internal stagnation, what seems to be quite clear is that working classes across the world, including the peasantries, are to be punished for whatever gains they had made under pressures of communism and national liberation struggles.

I was invited to speak on the theme of globalization, and I have therefore detained you for so long on the structure of the historical conditions on which this latest phase of imperialism has arisen. I have deliberately downplayed the economic structures of globalization because those are what get discussed the most and I therefore assume that my audience would know those arguments already. I have restricted myself, instead, to the political side of this process. As I said at the outset, this indeed *is* a radically new phase of imperialism. However, I now change my theme and offer some broad generalizations related to the issue of agriculture. I want to start with the observation that we shall make a very great mistake if we do not also see the fundamental, structural continuity between the earlier and the latter phases. Roots of the current crisis for the broad masses go deep into earlier histories, all the way into the colonial period when ours was almost wholly an agrarian society, and then into the very models of development—which is to say, structures of property relations—that the post-colonial states sought to enforce. To illustrate this point, let us take up a stark hypothesis.

Every national bourgeois regime that arose after decolonization across the predominantly agrarian continents of Asia and Africa had a stark choice of alignment. They could either align themselves with the peasantry at home and carry out an agrarian revolution, so as to lay the foundations for equitable property and production relations as prerequisites for a coherent form of independent industrial development with domestic bases. Or they could align themselves with imperialism, externally, in a relationship of dependence so as to stabilize a set of domestic property relations which were already favourable to them and launch a form of capitalist growth which was largely divorced from its agrarian hinterlands. It would be perennially dependent on external inputs in the shape of technology, aid, finance and even external markets, considering that a stagnant agrarian sector could not give it the required surpluses for industrial investment and the lack of an adequate home market would mean that import-substitution industrialization would soon give way to export-oriented production in search of foreign markets. Our historical experience is that faced with this choice of alignment with either the peasantry or with imperialism, virtually every national bourgeoisie betrayed the peasantry and aligned itself, mostly sooner rather than later, with imperialism.

This is a theme of great significance. Gramsci argues that the European bourgeoisie that went through the experience of the French Revolution got so frightened by the prospect of the peasantry carrying its own revolution to its logical end that no bourgeoisie was ever again to play a revolutionary role against the landholding classes. It follows, therefore, that the national-bourgeois regimes which arose in our continents at the crest of peasant mobilizations, as was indeed the case in India, were similarly determined that the peasantry be demobilized as quickly as possible, lest they march on in an uninterrupted motion from a merely anti-colonial moment in our national history to an authentically radical agrarian revolution. Be that as it may, in the agrarian economies of the larger former colonies, agrarian revolution was the only way out of imperialist dependence and aside from the unremitting violence practised by imperialism against the patriotic and progressive forces in the Third World, that lack of will to carry out an agrarian revolution at home lies very much at the heart of the defeat of the national-bourgeois project and the eventual acceptance of imperialist dictation and formation of neoliberal regimes by the local bourgeoisies.

What, precisely do I mean by 'agrarian revolution'. As a Marxian socialist, I would of course prefer an agrarian revolution of the socialist kind. As a student of history, however, I also know that creation of a socialist agrarian economy can hardly be the vocation of the kind of national-bourgeois states that arose in the aftermath of colonialism. Let me, therefore, limit myself to enumerating some of the kind of radical agrarian transformations I have in mind when I use the term 'agrarian revolution' in this context. The first is of course a kind of radical redistribution of land which breaks up the parasitic, large, feudal or semi-feudal estates in such a way as to address the very high incidence of landlessness in the countryside and create a *generalized* class of petty landowners in the countryside—not the 'elite farmer strategy' of the sort that the capitalist-landlord regime in India generally followed but something more akin to the land reforms pursued in Japan, South Korea and Taiwan which did make a transition to advanced capitalism not to speak of China, which did so in a socialist context.

Land redistribution is the basis but by no means a sufficient condition. The small farmer gains the impetus to increase the productivity of land when he owns it but he cannot by himself acquire the inputs, the skills, the conditions of health and education, the marketing mechanisms, the price stabilization, which are all required for advances in agriculture. All such can only come either from corporate investment in agriculture which is precluded by the land redistribution itself, or from the state or cooperative and collectivist modules in which small peasants pool their respectively meagre surpluses to build a large enough surplus to buy some of those inputs, which would still need to be supplemented by state expenditures on provision of health and education, infrastructural development, price support systems, and the like. Cooperatives themselves would need State credit in their initial phase. All of this of course raises the question of the nature of the state. What we lacked during the years immediately following independence was not the modest sums required for this support of the small peasantry, since India had actually inherited very handsome amounts of reserves from the British colonial state, thanks to the surpluses that the British Indian government had been accumulating during the Second World War in its transactions with the home government in Britain. Those reserves played a significant role in underwriting the first two five year plans and in getting the Indian government into the habit of neither taxing the landlords adequately nor collecting taxes from the capitalist class efficiently. Funds of that kind could have been used for revamping the rural structure, providing the poorer sections of the rural population with adequate inputs, infrastructural support, provision of health and education—factors that were deemed essential in Japan immediately after the Second World War and in South Korea and Taiwan from about the 1960s onward. What was lacking, rather, was the will to break-up landlordism all across India, and then to favour the small peasantry rather than the rich clusters among them.

Once agriculture has been modernized to any considerable degree, this modernization itself creates problems in the sphere of rural unemployment. Our historical experience in agriculture, from the earliest days of such transformations in England during the seventeenth and eighteenth centuries to the East Asian experience of the past 50 years, is that no robust capitalism, of the advanced type, arises unless agriculture itself is radically transformed. But it is also that rising productivities in agriculture necessarily lead to declining ratios of the population engaged in farm labour and proportionate rise in the aggregate number of the rural population *not* engaged in agriculture directly. In other words, labour force working the farms declines while numbers of non-farm labour increases. Some portion of this farm labour gets absorbed in related activities, such as marketing and transport, but there remains an absolute and rising number which must be absorbed elsewhere.

Patterns in countries that have made a full transition to capitalism are telling. In the first prolonged phase, the bulk of the surplus rural populations are absorbed in industry, so that the shifting balance of rural/urban population reflects fairly closely the agriculture/industry balance, with industrial labour force rising and agricultural labour force declining while what later come to be called 'services' absorbing relatively lesser proportion of the workforce. Only at the stage of mature, advanced capitalism do we find a situation in which the proportion of population directly engaged in agriculture declines absolutely (down to below 5 per cent in

Britain and the US for instance, and well below 20 per cent in any country that made that transition, right down to South Korea), the ratio of population in industry stabilizes and may even decline, while 'services' begin to absorb more and more of the workforce.

The more backward a country, the more likely it is that industry would absorb a miniscule proportion of the workforce, the organized sector would be relatively much smaller than the unorganized sector in the urban economy, vast bulk of the population would stay in the countryside while numbers engaged in farm work there decline and non-farm sources for earning a livelihood are sought by more and more people who nevertheless face at times insurmountable difficulties in finding such non-farm sources. India belongs to this category of the most wretched countries on earth, alongside other countries of South Asia and Sub-Saharan Africa. I might add that landlessness, uncertainty of tenure for the poor peasantry and pathetic conditions for agricultural work were elements that post-Independence India had inherited from the colonial period itself, in varying degrees from one region to another. Further declines in rural unemployment with the introduction of farm machinery and other features of the Green Revolution were grafted on top of that inheritance. This must be seen in the context of lop-sided development of capitalist agriculture which favoured the landlord class and the rich farmers, and in a context of low levels of industrialization which was concentrated, moreover, in and near the larger cities, favouring economies of scale in the public as well as private sectors, instead of widespread industrialization based on intermediate technologies and favouring towns and peri-urban townships organically related to the countryside. There was some chaotic industrialization in the provincial cities and towns as well but by no means sufficient for the bulk of non-farm labour to actually forego their rural habitats and move into the cities with secure, gainful employment. Thus it is that villages even in the most prosperous of our states, such as Punjab, are fast getting to be sites of uneasy coexistence between declining proportions of farm labour working in the village economy itself and the non-farm labour most of which goes elsewhere in search of employment or even petty entrepreneurship but has nowhere else to live. They are now, in a sense, refugees in their own villages.

Just one set of comparative figures should clarify what I mean. At the time of Independence in 1947, India was relatively the most industrialized among countries of Asia and Africa that were gaining Independence during that period, with the largest bourgeoisie and educated middle class. The Independence struggle itself was led by the likes of Gandhi and Nehru who never tired of promising liberation for the peasantry, while numerous historians, including leftist ones such as Bipan Chandra, have written copiously on the patriotic role played by some of the most powerful of the industrial houses in the Independence struggle. By contrast, Japan which had indeed been a major industrial power before the Second World War had been devastated by the war and had even sustained the only nuclear attack in human history. South Korea, with no prior industrial base, was similarly devastated by the Korean War and subsequent partition, beginning its recovery only around 1960. Taiwan, an island off the coast of Chinese mainland also had no prior industrial base when it was occupied by Chiang Kai-Shek's reactionary armies when they withdrew from the mainland toward the end of the Chinese Revolution in 1949.

With these comparative backgrounds in mind, let us look at what happened during the second half of the twentieth century. India reduced the proportion of population dependent on agriculture by 15 per cent over 44 years, from 76 per cent in 1950 to 61 per cent in 1994, while population growth over all those years of course meant that in absolute numbers many more people were dependent on agriculture at the end of that period (1994) than at the beginning (1950). By contrast, Japan reduced the number of farmers from 52.4 per cent in 1947 to 9 per cent in 1985 (over 37 years), Korea from 66 per cent in 1960 to 17 per cent in 1991 (31 years), and Taiwan from 56 per cent in 1952 to 9 per cent in 1998 (over 46 years). Even China, which was a much more backward and devastated country at the time of its revolution in 1949 than was India in 1947, and which developed its industrial and other modern sectors pretty much on its own well into the 1980s, nevertheless managed to reduce its agriculture-related population from 86 per cent in 1952 to 50 per cent in 1998. During those same years as given above for each country, industrial employment in India arose by 4 per cent, in China by 16, in Korea by 26. In Taiwan, where proportion of agricultural population had declined to below 10 per cent, making it comparable with Japan and the advanced Western countries, industrial employment had actually jumped by 25 per cent by 1985 but then declined by 4 per cent which moved to services. This expansion of services at the expense of industrial employment after a certain level of advanced industrialization has been achieved is also characteristic of advanced capitalism in general. Korea had displayed the same trend. India's peculiarity is that while industrial employment rose by mere 4 per cent over 44 years, the proportion of population in services doubled during the same years—a classic symptom of structural stagnation in agriculture as well as industry, regardless of the immense increase in output in some selected industrial sectors.

Japan, Korea and Taiwan did all this while integrating themselves closely with the US economy. China did so while it was still sequestering itself from the global economy altogether, and the trend has continued thereafter. India's agriculture-related population declined by 15 per cent between 1950 and 1994, the Chinese fell by 22 per cent in the years 1952–85 (considerably more and faster). It is roughly at this point that Chinese integration with the world capitalist economy actually started to take off, thanks to the Deng 'reforms' beginning in 1978.

I might add two more points. One, Indian performance in this regard is roughly comparable to that of Bangladesh but inferior to that of Pakistan or Sri Lanka, not to speak of Indonesia or Malaysia—even though India is now an industrial giant which none of those countries are. Second, India's inferiority, compared to the East Asian countries discussed, is indeed much more dramatic if we also take into account issues of health, housing, education, unemployment insurance, infrastructural facilities such as roads and transport, and other indices of security and well-being.

Rising rates of farm-related unemployment and increasing unviability of small farms are by no means peculiar to India but a structural feature of the generalization of capitalism in general, regardless of the patterns of landholdings. In the United States itself, which has set the trend for capitalization of agriculture throughout the world, in Western Europe as much as in Asia, farm population accounts for roughly 5 per cent of the population. Half of agricultural produce comes from 2 per cent of the farms which are all corporate-owned while family farms which are 73 per cent of all farms account for only 9 per cent of the production and derive an average of mere 14 per cent of their incomes from farming as such. This structure is sustained by a level of government support which gives subsidies to US farmers and agro-traders that are, per capita, 100 per cent higher than the average income of corn farmers in southern Philippines. Ninety-five per cent of all food in the US is manufactured and marketed by about a dozen corporations which are also the main exporters of US farm products, which in itself is no small aspect of globalization itself. By 1994, US agro-exports accounted for 36 per cent of wheat trade worldwide; 60 per cent of corn, barley, sorghum, and oats; 40 per cent of soyabeans; 17 per cent of rice; and 33 per cent of cotton. This export-orientation is then combined with globalized production and marketing. Cargill, the largest US based agro-firm, employs over 70,000 people worldwide in 60 countries in over 50 businesses including subsidiaries, and deals in a variety of products from grains and meat to peanuts, coffee, fruits and vegetables, seeds and fertilizers, and now an aggressive push into genetically modified crops.

Trends are set in the US but the same trends are then followed by Japanese and European multinational giants such as Mitsubishi, Nestle and so on. Some of the key features of this kind of globalized agriculture include corporate and contract farming, globalized trade in agro-goods, corporate-controlled production that is spread over multiple locations around the world, and increasing transformation of small farms into contractually-bound small producers for corporate buyers for whom even the inputs—seeds, fertilizers, machinery—are determined by the corporate contractor. Worldwide bankruptcy of small farms is the essential precondition for increasing dominance of agriculture by corporate entities. Advanced capitalist countries have resolved the problem of surplus rural populations by absorbing them into highly developed industries and services oriented toward national and global markets. In countries like India, where agricultural employment is declining absolutely and where industrial employment is essentially stagnant and the modern servicing sector is itself sluggish, no such solution is possible. In large parts of the Indian countryside as well as in the underbelly of our towns and cities most employments which absorb surplus rural populations, yield so meagre a livelihood that it is in reality a disguised form of unemployment. If you add to the unemployment that is indeed registered as such in national statistics, this immiseration covers by far the majority of the population.

A coherent transition to capitalism succeeds only to the extent that the increasing population segments rendered jobless in the farming sector are gainfully employed in the production of value in other sectors. Moreover, without rising incomes for the great majority of the population, expansion of the home market remains restricted and sluggish so that the more import-substitution industrialization succeeds and the capacity to produce high-value goods increases, the more a structural imperative arises for export-oriented production since rising production volumes cannot be absorbed at home. By now in India, we have completely abandoned the very idea that majority of the population needs to be absorbed in stable, gainful employment with enough and secure enough incomes for them to become buyers of industrial goods. This can be seen in two most radical steps the current UPA government has announced for the poor peasantry: the Rural Employment Guarantee Scheme (REGS) and the waiver of loans to the poorest of farmers. Days of the year covered in REGS are so few, the minimum wage fixed for it so low, and the implementation of the scheme by the state governments is so indifferent that even if we do not consider the graft and inefficiency associated with such schemes, it is not at all clear that the scheme can provide even bare subsistence over 365 days of the year. The much-brandished loan waiver covers too few of the farmers in distress. Also, it covers only government loans and, as it turns out, the governments will give the money to the banks anyway. This seems more a way to cover banks against their unrecoverable loans—than any real relief to the farmer whose conditions of life remain the same. Together, the two schemes amount at best to the legendary 'opium for the masses', i.e. temporary pain relief.

Meanwhile, the sheer magnitude of our population is a great boon for the capitalist class. The misery of the vast majority keeps the wage rate unbelievably low for the employed sectors whereas if even 40 per cent of the whole population can be turned into consumers of manufactured goods, India can have a home market as large as the US or the EU. With that kind of a home market available, and with no revolutionary challenge arising in the country, the state and the classes it represents need do absolutely nothing for the other 60 per cent. This is the real secret behind the kind of class structure we have: fabulous, First World levels of wealth for the top one per cent or so, adequate incomes for 30 per cent or so to buy manufactured goods, even on credit, and unspeakable conditions of life for the rest. Five of the world's richest billionaires are now of Indian origin, but if we apply the UNDP criterion for the poverty line, two-thirds or more of the population would fall below that line. India has more illiterates than all of Africa or Latin America. Half of the world's blind people live in India, resigned to their fate, while hundreds of thousands in Latin America are recovering their eyesight thanks to Cuban and Venezuelan projects. Again, the problem is not lack of funds or expertise but lack of national will.

This, in sum, is our encounter with what we now call globalization. One could go on explicating numerous other things: how globalization depressed prices for commodity prices in the Third World, leading to crisis of even rich farmer economies; how the Green Revolution onwards, it has forced huge sectors of the farm population to be dependent on capitalist inputs such as seeds and fertilizers, leading to unemployment as well as soil erosion; how the new corporate offensive, backed by multinationals, to penetrate and dominate the retail trade will not only ruin petty traders and hawkers but will depress farm prices while raising prices in retail stores; and how the dominance of retail will necessarily lead to a combination of corporate and contract farming. There are any number of such issues but this paper will become endless if we delve into all the features of the capitalist offensive that go by the name of globalization. If what has been said can provide some useful points for further discussion, one purpose shall have been served.

CHAPTER 4

India's Granary Devastated: Punjab under Neo-Liberal Economic Reforms

Utsa Patnaik

INTRODUCTION

Punjab had the benefit of being the last region in India to be colonized, it came under the British control as late as 1848. The region had a fair amount of public investment in canal irrigation, inducing a shift of population from the densely settled eastern districts (of the present-day Indian Punjab) to the west Punjab districts from the last decade of the nineteenth century, establishing the well-known canal colonies (Paustian, 1930). Wheat and cotton became important export crops. However, as the commercialization of agriculture proceeded at a fast pace it led to the sub-division of holdings and growth of tenancy in the region (Calvert, 1936). During the half century before Independence, the British India as a whole saw a nearly 30 per cent decline in food grains availability per head of population and an all-pervasive agrarian crisis as crop prices fell on global markets during the inter-war Depression. Punjab too was caught in the crisis of deepening indebtedness thus forcing the peasants to transfer lands to the creditors. Its growth slowed down, but it still continued to have the highest agricultural growth and hence experienced the lowest degree of decline in food grains availability in British India (Table 4.1), to the extent of 18 per cent compared to as much as 38 per cent decline in Bengal (Blyn, 1966).

The agricultural pre-eminence of a partitioned Punjab continued in the post-Independence period. After the mid-1960s crisis of

TABLE 4.1: DECLINE IN PER CAPITA FOOD GRAINS OUTPUT BY REGIONS IN BRITISH INDIA

Region	Years	Total decline	Decline per year
Greater Bengal	1901–41	38	1.18
United Provinces	1921–41	24	1.36
Madras	1916–41	30	1.40
Bombay-Sind	1916–41	26	1.21
Central Provinces	1921–41	19	1.05
Greater Punjab	1921–41	18	1.00
British India	1911–41	29	1.14

Source: Blyn, 1966, rearranged from Table 5.3 on p. 102.

agriculture when the new agricultural strategy was undertaken for raising growth by promoting the high yielding seed-fertilizer-water package, it were Punjab's landlords and farmers who adopted it with greatest enthusiasm. The profitability of agricultural production was rising, and to understand why this was the case, we have to recognize that the fiscal stance of the central government was strongly expansionary and development-oriented from the mid-1950s to the end-1980s. As public investment and expenditure rose fast under the central and the state government plans, and since these expenditures had a large wage component, through multiplier effect, there was a rapidly expanding demand in the country for necessities, especially food grains. Michael Kalecki in a well-known paper on the problems of financing development in a mixed economy had anticipated that inflationary pressures would be strong with planned development, since without substantial land reforms, agricultural output could not be expected to expand as fast as the demand was growing (Kalecki, 1972). The main financial constraint on non-inflationary growth could thus be looked upon as the problem of ensuring adequate supplies of food grains and raw materials from agriculture. All agricultural prices did indeed start rising from the late 1960s and the terms of trade (the ratio of agricultural prices to manufactured goods prices) shifted in favour of agriculture.

Punjab was well placed to respond to and meet the demands of an expanding internal market. The class structure in Punjab's villages was not as unequal and polarized as in the regions which had been longer under colonial rule, such as Bengal or Madras where land ownership was very highly concentrated and the majority of the rural population had been reduced to destitute labourers and poor peasants. No doubt there were landlords in Punjab who controlled a substantial part of the total cultivated land, but there was also a substantial class of middle and well-to-do peasant farmers able to undertake some investment, so the social base of investment was considerably broader than elsewhere.

As Table 4.2 shows, the output of food grains per head of population rose in Punjab by 113 per cent, or more than doubled, by the mid-1980s compared to the early 1960s. The entire northern region—Punjab, Haryana and Uttar Pradesh—registered a 65 per cent rise in per head food grains output. But it is a remarkable fact that not a single other state in India during the same period—other than West Bengal where this variable rose very marginally—escaped the experience of falling food grains output per head, continuing the same trend as in the pre-Independence period, though at a slower rate. In some states the fall was in the first period up to the 1970s with stabilization afterwards and in others the fall came in the second period. The observed overall all-India average rise in per capita food grains output was thus entirely the contribution of rapidly rising output in northern India and particularly of the success of the Green Revolution in Punjab which outweighed the decline elsewhere in the country where growth rates did not keep pace with population growth. Punjab, Haryana and Uttar Pradesh emerged as the granary of the country.

There was a social and political price that Punjab paid for the rapid growth of capitalist farming despite its prosperity. Landlords evicted lakhs of small and poor tenant farmers both in the 1960s and again in the 1970s as they switched to more profitable direct cultivation with hired labour. As a result Punjab was the only state in the country where the concentration of operated area actually increased. The share of wage-paid workers in the total workforce rose fast and that of cultivators declined. Not all the

TABLE 4.2: ANNUAL FOOD GRAINS OUTPUT IN KILOGRAMS, PER HEAD OF TOTAL STATE POPULATION 1960–2 TO 1984–6

State/Region	Per capita output		
	1960–2	1972–7	1984–6
NORTH			
Haryana & Punjab	313.5	454.00	734.9
Uttar Pradesh	184.5	176.92	42.8
Jammu & Kashmir and Himachal Pradesh	113.9	222.12	12.4
Region Average	204.6	234.70	337.2
(per cent change over 1960–2)	(14.7)	(64.8)	
EAST			
Assam	145.4	137.9	121.1
Bihar	158.6	140.0	136.9
Orissa	225.1	200.1	217.1
W. Bengal	147.5	151.0	154.6
Region Average	162.2	152.9	152.9
(per cent change over 1960–2)	(-5.7)	(-5.7)	
SOUTH			
Andhra Pradesh	180.8	175.3	161.5
Karnataka	161.6	185.0	154.3
Kerala	61.9	58.9	43.6
Tamil Nadu	160.9	146.6	134.1
Region Average	152.3	150.4	143.9
(per cent change over 1960–2)	(-1.25)	(-5.5)	
WEST-CENTRAL			
Rajasthan	241.1	199.8	180.4
Gujarat	103.5	95.2	95.2
Madhya Pradesh	273.9	231.4	237.2
Maharashtra	165.0	110.0	120.7
Region Average	198.6	158.3	160.3
(per cent change over 1960–2)	(-20.3)	(-19.8)	
ALL INDIA	178.9	172.0	192.1
(per cent change over 1960–2)	(-3.9)	(7.38)	

Source: Primary annual food grains output data from the *Economic Survey*, various years. Earlier published in U. Patnaik, 1994. Population data from *Census of India*, 1981.

displaced tenants could be absorbed; especially the small and middle peasants who by their culture and habits would not do demeaning work as hired labour, faced rising unemployment and falling living standards. This fuelled discontent which along with other factors, fed into the secessionist agitation that troubled Punjab for more than a decade. Substantial seasonal in-migration of mainly Bihari labour, however, sustained output expansion in Punjab through the worst years of militancy and State counter-violence.

Responding to the mid-1960s agricultural crisis, the central government formed the Agricultural Prices Commission (APC) and set up the Food Corporation of India (FCI), both in 1965. The former determined the minimum support prices for food grains and other crops, based on the current cost of production plus a reasonable return on the fixed outlays of the producers. The procurement price at which the FCI actually purchased from farmers was usually set a little higher. A network of fair price shops in the cities and villages distributed the procured grain and other goods at controlled prices. The period from the mid-1960s to 1975 was one of rapid food price inflation. The urban price-rise resistance movement of the early 1970s underscored the inflation-sensitivity of the vocal urban population and the importance of price stabilization. The scale of procurement operations of the FCI was expanded steadily and rapidly thereafter, and this was possible only on the basis of procuring the growing food grains surpluses from north India, particularly Punjab where wheat output was growing at 10 per cent annually. Nearly nine-tenths of wheat procurement and two-thirds of rice procurement came from the northern states of India. Traditionally, Punjab was not a large rice growing region nor was rice its staple crop, but when irrigated rice expanded rapidly, it was adopted as a commercial crop as more than 90 per cent was sold in the market.

The criticism that the Public Distribution System (PDS) had an urban bias was deflected by opening four lakh new shops mainly in rural areas during the 1980s. By this time the FCI was procuring up to one-sixth of the country's total gross food grains output and accounted for 40 to 45 per cent of total marketed supplies in the country, which went to urban areas and to rural areas of deficit states. The government's intervention in the food economy through procurement and distribution was thus very substantial and it automatically had a regulating effect on prices charged by the private traders, helping to stabilize food grains prices both for the consumers and for the growers. A central subsidy was provided for the purpose of meeting the costs of the FCI's storage, handling and transport of food grains from the northern surplus procurement areas to the deficit states. Without such a subsidy, the cost of food grains would have become inordinately high for consumers in the deficit areas, especially the distant parts of the southern and the north-eastern states of India.

The purpose of food subsidy was not primarily poverty alleviation, but simply to ensure that the central issue price of food grains from fair price shops would not vary widely but remain within a narrow band whether the shops were located in Delhi or Chennai or Dimapur. This reflected the obligation of a central government which had promoted regional specialization in agriculture, to ensure that the economic and political unity of the vast and far-flung Indian Union remains intact. It was only fair, for example, that Kerala whose large foreign exchange earnings from export crops and later, worker remittances benefited the Indian Union, should be ensured those food grains at affordable prices in whose production it was deficient. From the late 1970s, however, many states with high levels of poverty, in south India, Gujarat and West Bengal, started providing an additional subsidy out of their state budgets to cover local handling and distribution costs and reduce the issue price further for the benefit of the poor.

Today when the very existence of the PDS in the country is being threatened by reform policies, and we look back at the quarter century of PDS operation from 1966 to 1991, the conclusion is inescapable that despite all its problems and shortcomings, it has played a very important role in stabilizing crucial prices both for the growers of crops and for the consumers. For farmers who face the vagaries of the monsoon and hence face output instability which is outside their control, it is very important to be assured of some degree of price stability so that they can plan their investment and output without the fear of losing the money they have invested. Punjab's farmers benefited from the slow but steady rise in procurement price and the shift in the internal terms of trade towards agriculture during the period 1955 to 1970 when globally, the trend was the opposite—the terms of trade were shifting sharply against agriculture through a series of large fluctuations. Karshenas (1995: 89) estimates that farmers in India as a whole would have incurred an income loss of Rs. 27 billion if they had faced falling international terms of trade, but owing to the protection to agriculture and improving internal terms of trade, farmers actually gained Rs. 11 billion.

While Punjab's farmers benefited, they contributed equally to price stability and rising supplies for sustaining industrial growth, because they were indifferent to the sharp upward fluctuation in primary product prices on global markets in the early 1970s (see Figure 4.1). There was no outcry at that time that 'the Indian farmers

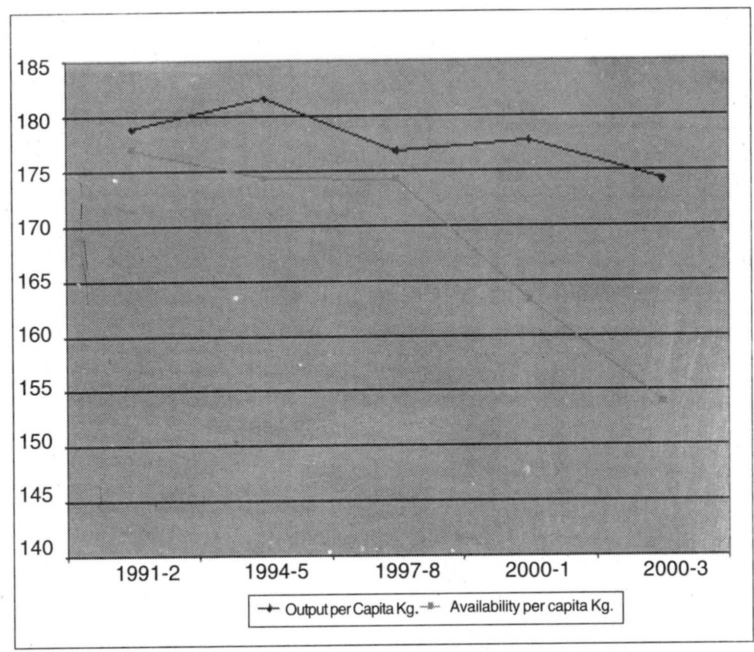

Source: Based on Table 4.7.

Figure 4.1: Food Grains Output and Availability per Capita per Annum, India, (Triennial average [1990–2 to 2001–3])

should get world prices'—an outcry which was to be mischievously engineered by pro-trade liberalization forces two decades later. Punjab's farmers sold their wheat surpluses happily to the FCI even though by 1974 the price they were getting was hardly one-third of the world price; they could not care less about the volatile world price. All that they wanted was a steady reasonable return, not large speculative windfall profits and they were satisfied with the steady annual rise in procurement price. If they had insisted on exporting at that time and maximizing their profits rather than selling to the FCI, the internal rate of food price inflation in India would have been much higher than it was, and real wages would have fallen even more sharply. Punjab's farmers were wise to be satisfied, because when world primary prices crashed sharply from the mid-1970s and continued their downward trend, they knew nothing of it, being completely insulated from the destabilizing effects of the global price volatility by the assured minimum price at which they continued to sell to the FCI.

How important is it to insulate farmers from declining global terms of trade and sudden price rise and price crash, especially in a developing economy where the majority of farmers are poor, can be judged from the present situation. With the unwise complete opening up of agricultural trade from the mid-1990s during a period when global prices were falling, lakhs of India's farmers have been plunged into deepening indebtedness and asset loss while many thousands have been driven to commit suicide. The number of suicides committed by farmers in Punjab too have been large, though officially admitted to be 2,116 only since 1997. But these numbers have been underestimated owing to non-reporting. The total number of officially admitted (reported in newspapers in May 2006) farmer suicides, including other areas like Andhra, Karnataka, Vidarbha and Kerala, is 15,000, but again this is an underestimation.

It is not only exposure to global price volatility, however, which underlies the present problems of Punjab's farmers. To understand these problems fully it is necessary to understand the effects of the deflationary economic reform policies on the demand for food grains within the country. (Here the antonym of 'deflationary policies' is not 'inflationary policies' because it is expenditure deflation in a situation of existing unemployment, which we are talking about.) Punjab's agricultural prosperity for a quarter century after 1965 was closely linked to its emergence as the pre-eminent supplier of food grains to the rest of India, when the central state could be termed a developmental state following expansionary policies. The argument of this paper is that Punjab's decline in prosperity today is the outcome of the sharp contraction of aggregate demand for food grains in India, especially in rural India, brought about by expenditure-contracting public policies advised by the international financial institutions and faithfully and foolishly implemented by successive governments over the last fifteen years. The fall in demand is non-voluntary and the result of falling purchasing power for the mass of the rural population, and it has resulted in deepening undernutrition in a number of states on the one hand and a corresponding loss of internal markets for Punjab, on the other. The effects of the loss of markets have been further aggravated by contraction in public development expenditure in Punjab itself which has raised the depth of rural poverty substantially as we will see in the last section.

The complete U-turn with regard to macroeconomic policies affecting the government's fiscal stance can be traced precisely to the election of 1991 which brought in a minority Congress government with Manmohan Singh serving as finance minister. A slew of policies were announced without any discussion, within a week of the government's formation—obviously they had been all worked out beforehand—including large cuts in public expenditure and in investment, sharp reduction in fertilizer subsidy, and a large devaluation of the rupee. The result was such a sharp reduction in the growth rate that the per capita GDP declined. There was easing up on the fiscal front in the run-up to the 1996 elections, and during the two years of the United Front government.

Expenditure deflating policies were resumed with a vengeance, however, under the National Democratic Alliance (NDA) government led by the Bharatiya Janata Party (BJP) from 1998 along with rapid dismantling of protection to agriculture. This produced a pervasive crisis of employment and livelihoods in rural India by 2004, as the next section documents. Expenditure deflating policies in the material productive sectors continue to be followed by the United Progressive Alliance (UPA) government which came to power after the May 2004 elections, as rigorously as before—indeed the target of reducing the fiscal deficit was incorporated in the Fiscal Responsibility and Budgetary Management Bill, the enactment of which was one of the first measures of the UPA.

Punjab's agrarian problems thus cannot be understood by focusing on Punjab alone. It requires contextualization of the problem and situating that within the entire country's changed macroeconomic policies and the systematic operation of deflation in the material productive sectors. The next section discusses the meaning and effects of deflationary policies.

WHY DEFLATIONARY POLICIES?

Expenditure-reducing macroeconomic policies are strongly favoured by international and domestic financial interest groups who are quite obsessive about controlling inflation, and would prefer to see even an economy with a high rate of unemployment, growing slowly and raising unemployment further, rather than risk the remotest possibility of price rise owing to expansionary policies reducing unemployment. Of course, excessive price inflation can and should be controlled by more benign and virtuous methods like undertaking radical redistributive land reforms, for one, to release the rapid growth of productive forces in agriculture which would raise supplies as fast or faster than aggregate demand is expanding within a development-oriented strategy; and by raising the rate of public investment in agriculture. International creditors however have uniformly advised developing countries to take the vicious route of demand deflation to prevent inflation, even in situations of high unemployment as in India today.

The interests of financiers—those who make money exclusively by dealing in money—have always been opposed to the interests of those engaged directly in production. The latter would like to see low real interest rates and expanding mass purchasing power. Financiers on the contrary, wish to maintain high real values of their financial assets and high real interest rates (inflation would erode both), and are happy with bouts of asset deflation in developing countries so that these assets can be snapped up at low prices by their corporations. Their insensate and obsessive fear of inflation can be seen in the expenditure-reducing policies advised uniformly by the International Monetary Fund (IMF) to 78 developing countries in the 1980s and summarized in Table 4.3 from an IMF study. The first three policies—restraint on central government expenditure, limits on credit expansion, and reduction of budget deficit to GDP ratio—add up together to a strongly deflationary package and all three were implemented at the same time by four-fifths of the concerned countries, while two-thirds capped wages and over half devalued their currency.

The results of deflationary policies of the decade up to the mid-1980s have been documented as sharp decline in rates of investment in both capital formation and in the social sectors, leading to reduced or negative GDP growth and negative impact on the human development indicators (see in particular Cornia et al., 1987). A number of studies since then have confirmed the adverse impact and have argued for expansionary policies.[1]

India has been following exactly the same deflationary package of policies since 1991, whose impact has been especially severe in India's agricultural sector which saw sharp reduction in public planned development expenditures in rural areas. Development

TABLE 4.3: POLICIES FOLLOWED BY 78 COUNTRIES UNDER FUND-GUIDED ECONOMIC REFORMS

	Percentage of total number of countries implementing policy
Restraint on central government expenditure	91
Limits on credit expansion	99
Reduction in ratio of budget deficit to GDP	83
Wage restraint	65
Exchange rate policy	54

Source: IMF study quoted in Cornia et al., *Adjustment with a Human Face*, 1987, vol. 1, p. 11.

expenditure in rural areas traditionally includes agriculture, rural development, irrigation and flood control—all vital for maintaining output—to which we add also the outlays on special areas programmes, and village and small-scale industry to define overall 'Rural Development Expenditures' or RDE. The employment-generating programmes had assumed special importance from the drought year 1987 onwards.

Over the Seventh Plan period marking the pre-reforms phase, from 1985 to 1990, Rs. 51,000 crore was spent on rural development, amounting to almost 4 per cent of Net National Product (NNP). By the mid-1990s, annual spending on rural development was down to 2.6 per cent of NNP, and further declines took place to only 1.9 per cent by 2000–1 (see Table 4.4). The real development expenditures rose less fast than population, producing per capita decline. I estimate that in constant 1993–4 prices about Rs. 30,000 crore less was being spent by the end-decade year 1999–2000, compared to the beginning, 1990–1. Every Rs. 100 less of spending results in income loss of a multiple of this, Rs. 400–500 depending on the value of the Keynesian multiplier which is given

TABLE 4.4: REDUCTION IN RURAL DEVELOPMENT EXPENDITURES UNDER ECONOMIC REFORMS (SELECTED YEARS) INDIA 1985–90 TO 2000–1

	1985–90 average	1993–4	1995–6	1997–8	2000–1
RDE as per cent of NNP	3.8	2.8	2.6	2.3	1.9
Above plus infrastructure	11.1	8.4	6.9	6.4	5.8

Source: Government of India, Ministry of Finance, annual *Economic Survey*, for years 2001–2 to 2003–4, Appendix Table S-44. 'RDE' here are the plan outlays of centre and states under the five heads of agriculture, rural development, irrigation and flood control, special areas programmes, and village and small-scale industry. Infrastructure includes all energy and transport including urban. Calculated from current values of expenditure and NNP at factor cost.

by the inverse of the marginal propensity to save. A crude point-to-point comparison would suggest an annual income loss for rural producers between Rs. 1,20,000 to Rs. 1,50,000 crore assuming a multiplier value between 4 and 5. Actual income loss would have been greater taking the cumulative losses over successive years. This harsh contractionary policy had nothing to do with any objective resource constraint but simply reflected the deflationary dogmas of the Bretton Woods Institutions (BWI),[2] which were internalized and sought to be justified by the Indian government.

There is no economic rationale for believing that 'public investment crowds out private investment' which is the common argument put forward for reducing the state's role in rural development. Precisely the contrary has been shown to hold for certain types of investment essential for an irrigation-dependent agriculture like India's such as irrigation projects of all types. Private tube well investment is profitable only where the water table remains high owing to seepage from state-built canal irrigation systems, and where community-integrated watershed management (planting trees and using check-dams) is encouraged with state help. Private overexploitation of ground water has now reached a crisis point in many states in India, with the water table falling rapidly and with even the richest farmers unable to reach water after investing heavily in deep bore wells and submersible pumps. Other infrastructure investment such as rural power projects, roads, bridges, school buildings, clinics and so on, are never undertaken by private investors but are vital for stimulating development and providing livelihoods both directly to those employed in building them and through the important multiplier effects on employment and incomes, of the increased wage incomes being spent on simple consumer goods and services within the villages. The market for machine-made textiles and other goods also thereby expand.

The agricultural universities had earlier played a major role in developing and helping to disseminate new crop varieties. The cut in funding for research in these universities, by affecting the search for better rain-fed crop varieties, has also contributed to the deceleration in the growth of yields. With increasing use of land for commercial and residential purposes, the gross sown area in India has remained static since 1991. So it is only through yield rise that output growth can be maintained and it is here that the failure is evident.

The net result of the unwise cut-back of public investment and in RDE has been a slowing of output growth—both food grains and non-food grains growth rates almost halved during the 1990s compared to the pre-reform 1980s, and both have fallen below the population growth rate even though this too is slowing down (Table 4.5). This led to declining per capita output during the 1990s, for the first time since the mid-1960s agricultural crisis. The latter, however, had been short-lived, whereas per head agricultural output continues to fall even more rapidly at present even after a decade.

The situation in the first five years of the twenty-first century has been even more grim than in the 1990s as shown in Table 4.5. Food grains growth rate is down to 0.14 per cent, virtually the same negligible rate amounting to stagnation as the long-term rate under colonialism in the fifty years before Independence, which was 0.11 per cent (Blyn, 1966). I had pointed out a decade ago (Patnaik, 1996) that since our land is limited and is not a reproducible resource, the only way that output growth can be maintained is through investment, so that trade liberalization

TABLE 4.5: DECELERATING GROWTH RATES OF AGRICULTURAL OUTPUT

Period	Food grains	Non-food grains	All crops	Population
1980–1 to 1989–90	2.85	3.77	3.19	2.1
1990–1 to 2000–1	1.66	1.86	1.73	1.9
2000–1 to 2005–6	0.14	n.a	n.a	1.8

Source: Government of India, Ministry of Finance, *Economic Survey, 2001–02,* p. 189 for first two lines. Last line calculated taking initial 2000–1 and terminal (advance estimate) 2005–6. Food grains output figures from *Economic Survey 2005-06* (both were near-normal rainfall years). Note that the fall in output growth is much steeper than slowing down of population growth, hence per head output is falling.

entailing unregulated exports from agriculture was bound to lead to decline in basic food grains output in a BWI-guided deflationary regime where overall investment rates in agriculture were not rising. Further, I had pointed out that the supply slowdown is complemented and reinforced by expenditure-deflating policies which reduce the growth of mass demand for basic food grains, viz., through a reduction in aggregate demand owing to rise in unemployment and loss of incomes. This serves the interests of the advanced countries which wish to see our food grains producing land being diverted to the luxury crops which serve to fill their supermarket shelves and diversify further the already diversified consumption basket of their rich consumers.

It is clear that the combination of decline in RDE and the near-halving of agricultural growth have indeed produced a major crisis of rising unemployment. There is both fast growing open unemployment and fall in number of days employed of the rural workforce during the economic reforms period. Even with constant labour coefficients (labour days used per unit of crop output) a near halving of employment growth was to be expected given the decline in crop output growth, but the decline in jobs has been even more as mechanization, especially of harvesting, and use of weedicides has led to falling labour coefficients. Further, the rural non-farm employment growth, which was robust in the 1980s owing to reasonably high state spending, declined in the 1990s. The ratio of labour force to population, or the participation rate, has fallen reflecting difficulty of finding work. The ratio of workforce to labour force has also declined because open unemployment has been growing at over 5 per cent annually (Table 4.6). The elasticity of employment with respect to agricultural output was 0.7 between 1983 and 1993–4 but has declined to 0.1 or virtually zero, in the reforms period 1993–4 to 1999–2000.

Our economists and policy-makers have a completely incorrect understanding in which they refuse to see that in rural areas the demand curve as a whole has shifted down. They are focused only on the price variable, implicitly assuming an unchanged demand curve. They have wrongly advocated reduction in the minimum support price (MSP) for farmers, incorrectly blaming mounting

TABLE 4.6: EMPLOYMENT DECLINE IN RURAL INDIA

	1983	1993–4	1999–2000	Growth per annum, % 1983 to 1993–4	1993–4 to 1999–2000
RURAL					
Population (mn.)	546.6	658.8	727.5	1.79	1.67
Labour force (mn.)	204.2	255.4	270.4	2.15	0.96
Workforce (mn.)	187.9	241.0	250.9	2.40	0.67
Unemployed (mn.) (2–3)	16.3	14.4	19.5	–1.19	5.26

Source: Government of India, Ministry of Finance, *Economic Survey 2002–03*, p. 218. Rural employment refers to all productive activities in rural areas including secondary and tertiary in addition to primary sector activities.

public food stocks up to 2002 on 'too high' MSP. There is not a word on demand deflation in their writings, though rising unemployment and falling purchasing power have been glaring realities for several years. There was indeed a mindless freeze by the government of the MSP even though input costs were rising, which only helped to further reduce the incomes of farmers and deepen the downward shift in aggregate demand. The blind and unreasoning advocacy by our policy makers in favour of 'diversification' and the entry of food business Trans National Corporations (TNCs) at the expense of steadily declining food staples, which can only benefit advanced countries at the expense of our own poor, is quite inexplicable. Do they still fail to see the obvious? Are thousands of farmer suicides not enough for them? Will it require a large-scale famine in the country before these policies are reversed?

Let no-one imagine that unemployed rural workers are migrating and finding employment in industry: there have also been massive job losses in manufacturing during the reform period. The share of the secondary sector in GDP has stagnated while the share in the workforce has fallen from 29 to around 22 per cent during the 1990s. In short, India has witnessed deindustrialization. The agricultural depression has reduced the share of agriculture in GDP from about a third at the beginning of the 1990s to just over a fifth a decade later, but the labour force and population dependent on agriculture have hardly fallen reflecting decline in per head incomes. Thus, the material productive sectors have declined and the only sector which has ballooned in an abnormal manner[3] is the tertiary or service sector which now accounts for over half of the GDP.

PUNJAB AGRICULTURE AND FOOD GRAINS ABSORPTION IN INDIA

Food grains absorption in India today has reached a historic low as a result of the massive decline in purchasing power, especially in villages, owing to the combination of rising unemployment, rising input and credit costs for farmers and exposure to global price declines from the mid-1990s. The loss of purchasing power is pervasive in rural India affecting both the 158 million wage-dependent workers as well as the 120 million cultivating workers and their families. Targeting the food subsidy from 1997–8 by restricting supply of cheaper grains to those officially identified as 'below the poverty line' has also added to the institutional denial of affordable food grains to the poor, not merely owing to mistakes of wrong exclusion from the set of the officially poor, but also owing to the gross official underestimation of the numbers in poverty, discussed at the end of this paper. The off-take of food grains from the PDS declined from 20 million tonnes in 1991 to barely 13 million tonnes by 2001, whereas normally it should have risen as the population rose (Swaminathan et al., 2002). The poor were duped by the government's misconceived attempt to reduce food subsidy through a steep hike in the issue price of food grains.

Independent India had seen forty years of slow but successful effort to raise per head food grains absorption from an abysmally low 152 kg average during the First Plan (1950–5) to the maximum level so far, 181 kg by the triennium centred on 1992–3. These forty years of effort have been wiped out in a mere decade of mass income-deflating policies and by July 2002 a mountain of unsold public food grains stocks of 63 million tonnes had built up, 40 million tonnes in excess of the July buffer norms.[4] This occurred owing to the steep decline in food grains absorption (availability) to 153 kg per head by the triennium ending in 2002–3 (see Table 4.7)—the same level as during the First Plan. This is lower than the average level of 157 kg food grain availability during 1937–41, in colonial India—which had declined further to reach 136 kg by 1946. In a series of earlier papers I have discussed the issues of long-term food grains availability trends, demand deflation during the 1990s and abnormal food stocks (for a detailed discussion see Patnaik, 2003a; 2003b; 2004; 2005).

India has exported record volumes of wheat and rice during the last six years, and its share in global exports of rice and wheat has risen quite noticeably. Despite the drastic slowing down of output growth noted in Table 4.3, India exported 22 million tonnes of food grains during the two years 2002 and 2003 (Bhalla, 2005), and the share of grain exports in total exports had risen from under one-fifth to almost a quarter. There is higher global trade integration reflected in rising trade-GDP ratio. During the severe drought year starting from monsoon 2002, despite grain output being 30 million tonnes lower than in the previous year, from June 2002 to November 2003, a total of 17 million tonnes of food grains were exported by the then NDA government. Superficially it looks as though policies of trade liberalization have 'worked'.

However, the crucial fact which is suppressed in official publications and in the writings of pro-reform economists, and this is true even after the elections and the change in government, is that the vastly increased grain exports have been coming out of more and more empty stomachs as millions of rural labourers and farmers have suffered job loss and income decline. The government and the majority of economists continue to pretend that the exports represent a surplus over what people voluntarily wish to consume—that internal demand for food grains has fallen because people are voluntarily adopting diets away from food grains towards higher value foods. Official publications like the *Economic Survey* and the Reserve Bank of India's *Report on Currency and Finance* articulate the fairy tale of voluntary diversification but do not raise or answer

TABLE 4.7: SUMMARY OF ANNUAL PER CAPITA FOOD GRAINS OUTPUT AND AVAILABILITY IN INDIA IN THE 1990s (THREE YEAR AVERAGE)

Three-year availability periods ending in	Average per head population (million gm/day)	Net output per head				Net food grains (kg/year)	
		cereals (kg)	food grains (kg)	cereals (kg)	pulses (kg)		
1991–2	850.70	163.43	178.77	162.8	14.2	177.0	485
1994–5	901.02	166.74	181.59	160.8	13.5	174.3	478
1997–8	953.07	162.98	176.81	161.6	12.6	174.2	477
2000–1	1,008.14	164.84	177.71	151.7	11.5	163.2	447
2002–3#	1,050.67	153.85	164.09	142.9	10.1	153.0	419
Individual year							
2003–4*	1,087.6	158.33	170.83	n.a.	n.a.	n.a.	n.a.
2004–5*	1,107.0	151.21	162.35	n.a.	n.a.	n.a.	n.a.
2005–6	1,126.0		159.76	n.a.	n.a.	n.a.	n.a.
Change in per capita availability of food grains %							
Triennium ending 1991–2 to triennium ending 1997–8				–1.6			
Triennium ending 1997–8 to triennium ending 2002–3				–12.2			
Total change, 1991–2 to 2002–3					–13.6		

Source: For output, trade and stocks, Reserve Bank of India, *Report on Currency and Finance*, various years; Government of India, Ministry of Finance, *Economic Survey*, various years. For population, from the census population totals for 1991 and 2001, annual growth rate of 1.89 per cent has been derived and used to interpolate for inter-censal years. Before 1991 and from 2001 onwards, the population figures given in the *Economic Survey* have been used.

* Indicates provisional.

\# This triennium should begin with 2001–2 and end with 2003–4. The absence so far of published trade and stocks change data for 2002–3 means availability cannot yet be calculated for that year. So we have taken the three year average of 2000–1, 2001–2, and 2002–3. The first year of this triennium thus overlaps the last year of the previous one.

the question: How can people suffering employment loss and facing unprecedented crop price declines be inferred to be better off and be voluntarily reducing cereals demand? And how is it that the current reduced level of total absorption of food grains per head of about 155 kg per annum is not seen in any country except the least developed and the Sub-Saharan African countries? For comparison, China absorbed 325 kg of food grains per capita annually in the mid-1990s when Indian absorption was less than 200 kg The USA absorbed over 850 kg of grains per capita annually.

The observed falling share of food expenditure in total expenditure for almost every expenditure group is officially cited as proof of every income segment including the poorest diversifying their diets and becoming better off, and seems to have persuaded some academics. No attention is paid to steadily falling average calorie intake in rural India as 'diversification' proceeds. The argument is quite fallacious and is based on a simple confusion between the necessary and sufficient conditions for improvement.

A falling share of food expenditure in total expenditure, as well as a falling share of grain expenditure in food expenditure, are necessary, but not sufficient indices of the consumer becoming better off, particularly when we are considering not a rich population in an advanced country, but a population already at a low standard of living. The share of food spending from the total spending can fall and is actually observed to fall when people are getting worse off because their real income is constant or falling. Owing to greater monetization of the economy and higher cost of utilities they are forced to spend more on the bare minimum of non-food essentials like fuel, transport and medical costs, and therefore are forced to cut back on food expenditure. Thus even when 'real income' is unchanged over time, some food expenditure has to be sacrificed to buy fuel (which was partly freely gathered and gleaned from common property resources earlier). Fuel is jointly demanded with food grains since no-one can eat raw grains. The poor also have to incur higher transport costs than before in search of work, incur higher medical costs, and so on.

Since the overwhelmingly large part of food expenditure itself is on staple grains, it is this which falls when food expenditure is cut.

Data for Sub-Saharan African countries show dietary 'diversification' as per capita income declines. We observe a falling share of calories from cereals and rising share from animal products, but this is because expenditure is falling and people are forced to cut back on already very low food spending, so there is an enforced large decline in cereal intake and absolute calorie intake per day is seen to decline quite steeply (Food Balance Sheets of the FAO, quoted earlier in Patnaik, 2003b). In effect, a Sub-Saharan Africa already exists in rural India today.

The availability of food grains, following the definition used by the Government of India, is the net output of food grains (namely, gross output minus one-eighth on account of seed, feed and wastage) plus net imports and minus net addition to public stocks. As Table 4.7 shows, in India as a whole the net output of food grains has fallen by about 15 kg per head, from 179 kg annual average before reforms started to 164 kg in the period ending 2002–3. But the availability, or absorption of food grains has fallen much more, by as much as 24 kg over the same period. The average Indian family was absorbing almost 100 kg less of food grains annually and four-fifths of the fall has come after 1998.

This is an enormous fall and very abnormal in any economy where per capita income is rising. Never in the past in our own country has food grains absorption declined as a trend when per capita income has risen as a trend. This food grains availability figure, being calculated directly from output adjusted only for trade and change in stocks, represents the amount absorbed for all possible uses in the economy—directly for consumption by humans, indirectly as feed for producing livestock products (milk, butter, eggs, poultry, meat, etc.), industrial use for producing commercial starch, alcohol, etc., and lastly processing into consumer items for a mainly urban market.

The only explanation for this massive fall when average per capita income has been rising is a great increase, not just in income inequality, but of income inequality of a particular type—absolute decline in the purchasing power of the majority of the population especially the rural population, combined with steep rise in the purchasing power of a minority. The latter minority is voluntarily diversifying diets and absorbing higher amounts of food grains (remember this absorption is not just cereal consumption directly but also animal products, alcohol, and so on). But this rise is more than outweighed by the steep fall in food grains absorption by the vast majority of the population suffering absolute decline from already low levels. They are sinking deeper into undernutrition. Those without the ability to reduce intake further are driven to indebtedness and sale of assets including land, in order to stay alive.

This steep fall in purchasing power and in food grains absorption has meant trouble for Punjab's farmers who have correspondingly lost a large internal market for their wheat and rice output. It is easy enough to calculate from Table 4.7, that if absorption per capita which was 177 kg in the triennium ending in 1992–3, had remained constant at this level instead of dropping to 153 kg by the last triennium ending 2002–3, then the 24 kg per head difference would have meant, multiplying by the 2002–3 population, an additional annual demand in the country for 26.5 m. tonnes of food grains. This is the minimal estimate of the loss of markets for the farmers of the northern Indian states, and the extent of hunger that the already poor have been forced to endure in the country.

In fact, with rising per head GDP over this period, if absolute decline in incomes had not taken place for the majority, and income distribution had remained unchanged, then the elasticity of food grains demand with respect to income would have remained unchanged. The absorption of food grains should have registered a rise and reached about 185 kg by 2003–4, projecting the pre-reform rate of rise of absorption. The estimated loss of markets is much greater if we consider this scenario rather than taking the early 1990s level as constant. The demand for food grains should have been 33 million tonnes higher than it is at present. With an expanding market, farmers would also have had the incentive to sow more area and produce more.

Instead, output of wheat plus rice has been stagnant in Punjab at 24–5 million tonnes since 1999–2000. This unfortunate situation has been the result of incorrect public policy based upon an incorrect analysis of the problem. Having itself created a situation of declining incomes and hence declining sales from the PDS leading to build-up of massive stocks and rising subsidy bill merely to meet the cost of holding the stocks, the government has then turned around and incorrectly said that 'output is too high relative to demand because MSP is too high'. The MSP was frozen at the same level for two years despite input costs rising, and procurement operations were run down, with the consequence that the north Indian farmers' incentive to produce food grains was reduced and output growth decelerated even faster to reach the negligible, virtual stagnation rate of 0.14 per cent over the five years ending in 2005–6.

This situation of market loss is likely to change only if there is a serious implementation of the rural employment guarantee scheme by the UPA government led by the Congress. The National Rural Employment Guarantee Act was passed and implementation started from 1 February 2006. Within the first month, four million persons registered seeking work under the Act and as knowledge regarding it spreads more will register. With an adequate expansion of funding and increasing employment under the Act, the demand for food grains is bound to register a sharp rise in future. Wages going to those employed will not only generate demand directly but will have multiplier effects in generating incomes in the hands of artisans and suppliers of simple manufactures. If this increased food grains demand in future is to benefit Punjab's farmers, then it must be ensured that procurement by public agencies in India is continued and expanded, and the space is not given to private companies. The increased supply should not be cornered by countries like Australia which was already exporting four million tonnes of wheat to India in 2006.

INDEBTEDNESS AND POVERTY IN RURAL PUNJAB AND DEFLATIONARY POLICIES

While farmer suicides, mainly in the cotton belt but also elsewhere, have been a tragic expression of the crisis of Punjab's agriculture brought about by neo-liberal policies of exposure of farmers to global price decline, the suicides represent only the tip of an iceberg. There is a deepening spiral of indebtedness leading to loss of assets to creditors and pauperization of the small peasantry. Landlessness has increased and in the worst-affected districts of Punjab, substantial land transfers have taken place to non-cultivating creditors like the *arhtiyas*, traders who double as suppliers of inputs on credit to farmers. Event the hitherto well-to-do farmers have faced such a squeeze on profits that there is a tendency to lease out land once more rather than cultivate directly.

The National Sample Survey Organization (NSSO) has been carrying out a 'situation assessment survey of farmers', and report number 498 called *Indebtedness of Farmer Households* (59th Round, January–December 2003) shows that in Punjab, 52 per cent of the total loan amount was from high-cost non-institutional sources in which professional and agriculturist moneylenders plus traders were prominent, accounting for 45 per cent of this total loan amount.[5] The average amount of outstanding loan per farmer in Punjab was highest in India at Rs. 41,576 with the average for farmers who were not from SC, ST and OBC background, standing at Rs. 66,147. The next highest level of outstanding debt per farmer was Rs. 33,907 in Kerala. Thus Punjab agriculture is characterized by the highest level of outstanding debt in India, the average for India being Rs. 12,585 only (ibid.: 26).

However, the high level of debt in Punjab, was also true during the successful Green Revolution years. It is not debt as such which is the problem—modern high-value agriculture cannot run without loans taken by the producer, any more than any other small business can. Rather, the problem arises when credit becomes unrepayable because misguided neo-liberal policies ensure that input prices rise even as output prices are allowed to fall, and the farmer is driven to the usurious moneylender because priority sector lending is redefined in such a manner that institutional funds no longer go to agriculture.

While the Planning Commission has been claiming low levels of rural poverty (37 per cent for all-India in 1993, using NSS 50th Round data) and reduction in rural poverty by 1999–2000 in all states (to 27 per cent using 55th Round data), these are spurious claims. The reality is that poverty is very high and has been increasing in all except four states in India. The official estimates are incorrect because the rural nutrition norm has been abandoned owing to the procedure followed after the initial 1973–4 estimate of poverty line. This procedure has been simply to update this base year poverty line by using a price index. The base year by now is over three decades in the past and the price-adjusted poverty line permitted less than 1,900 calories to be accessed at the all-India level by 1999–2000, while at Punjab's official poverty line less than 1,720 calories could be accessed.

It is not a logically valid procedure to estimate poverty or compare change in poverty over time when the nutrition standard is being lowered. The correct method is to directly ascertain from the current NSS consumer expenditure data, what is the monthly per capita expenditure required to access the nutrition norm and what percentage of the population is below it. Poverty depth can also be directly determined by looking at the percentage of population spending less than the amount required to access lower energy intakes, like 2,100 and 1,800 calories. Table 4.8 summarizes the estimates for Punjab along with other states in north India (the corresponding data for all states is available in Patnaik, 2006).

The share of Punjab's rural population unable to access the nutrition norm of 2,400 calories per day was stable during 1977–8 and 1983 at about 41 per cent but it rose to 52.5 per cent by 1993–4 and to nearly 59 per cent by 1999–2000. The depth of poverty also increased, for the share of population with energy intake below 2,100 calories rose from 24 to 36 per cent between 1983 and 1999–2000. Thus lakhs of people have been pushed down into a lower nutritional status. Both Haryana and Uttar Pradesh, however, slightly improved their position over the period when Punjab registered deterioration. The share of population below 1,800 calories intake remained stable in all three states at 11 per cent.

The official estimate of 'rural poverty' in Punjab was only 12 per cent in 1993–4, declining to a negligible 6.4 per cent in 1999–2000. The only reason for these totally unrealistic numbers and incorrect claim of poverty reduction is that the relevant official

TABLE 4.8: RURAL POVERTY IN PUNJAB IN THE CONTEXT OF NORTH INDIA

State	Percentage of population spending less than amount required for obtaining specified intake of calories per day			
	1977–8	1983	1993–4	1999–2000
PUNJAB				
Calories per day				
<2,400	41.0	40.5	52.5	58.5
<2,100	20.5	24.0	30.0	36.0
<1,800	7.00	11.5	11.0	11.0
HARYANA				
Calories per day				
<2,400	n.a	53.0	55.0	47.5
<2,100	n.a	27.5	34.0	30.5
<1,800	n.a	08.5	11.5	11.0
UTTAR PRADESH				
Calories per day				
<2,400	48.0	57.5	65.5	61.5
<2,100	28.5	40.0	38.5	37.5
<1,800	11.0	14.0	13.5	11.0

Source: Calculated by author from NSS surveys on Consumer Expenditure for Rounds 32, 38, 50 and 55. Note that the calorie intake data for Haryana are not available for 1977–8. The calorie intake data are not available for any state for 1987–8. Calculations for all states, available in Patnaik, 2005 and 2006.

poverty lines are so absurdly low that a maximum of only 1,825 calories per diem could be accessed at this line in 1993–4, which dropped further to only 1,710 calories accessible at the 1999–2000 official poverty line. A real situation of high and increasing poverty in Punjab is converted to a low and declining official level of poverty solely because the official poverty lines, derived by applying a price index to a base year which is over three decades in the past, cannot capture the actual current cost of accessing the nutrition norm.

The method is logically incorrect because it ends up counting the poor as those below a steadily declining consumption standard over time. By 2005–6 the calorie intake corresponding to the official poverty line in Punjab will be about 1,600 only. A more detailed discussion of the issues and critique of the official estimates is available elsewhere (Patnaik, 2005 and 2006).

The seriousness of the food security problem and the general agrarian crisis both at the national level and in Punjab, evidently is not understood either by the central or the state government. The fact that there is both severe deficiency of aggregate demand in rural India as a whole, reflected in fall in food grains absorption, leading in turn to lowered production incentives and hence lowered growth rate of output owing to the wrong signals sent by government by way of freeze in MSP and inadequate procurement price, is not appreciated. The correct measure on the supply side is to take steps to raise food grains production, and on the demand side to raise aggregate demand through an effective implementation of the employment guarantee scheme.

But the policy pronouncements of the central and state government officials comprise precisely the opposite understanding. They repeatedly state that Punjab should 'diversify' production towards export crops under the aegis of the foreign TNCs engaged in the global food chain and suggest corporatization of agriculture as the solution. They have evidently never heard of falling international terms of trade for primary products and nor do they seem to be aware that all international bodies predict continuing flatness in non-oil and non-mineral primary prices up to 2010 at least.

The Punjab government has been cutting rural development expenditures despite agricultural depression in the state and thereby aggravating the existing problem. The Reserve Bank of India's data on state finances show that the total of plan and non-plan expenditures on rural development in Punjab, which was Rs. 1,764 crore in 2003–4 was cut to Rs. 1,258 crore by 2004–5, all in current prices (RBI, 2006), which means that the real decline is even greater. Punjab has already gone quite a long way towards facilitating the entry of foreign TNCs into its agrarian sector both through contracting systems and through direct acquisition of land.

The 'solution' of corporatization however is no solution but merely a gross aggravation of the disease. It will further deepen food insecurity and poverty by leading to a further collapse of food grains output growth, diverting more resources to export crops in a situation of stagnant or falling world prices for these crops. The TNCs may give good purchase price for the first few years to entice Punjab's farmers into their web, but they will then squeeze the farmers hard in order to maximize their own profits and reduce them to a situation of debt-bondage. The experience of other developing countries in Latin America and Sub-Saharan Africa, which have already gone in for corporate agriculture over the last two decades, aimed at supplying the supermarkets of northern countries has been far from happy. This experience should be studied before rushing into further unwise policy measures undermining our food security. A large literature now exists investigating the impact of contract farming under corporate control on farmers' livelihood, and the overwhelming evidence is that their livelihood becomes more precarious.

The real solution to the pervasive agrarian crisis thus lies in measures to improve productivity through increased public investment particularly in raising the output of the basic food grains required by the mass of our population both for direct consumption as well as for conversion to animal products. This has to be combined with a large expansion in rural development expenditures including a substantial effort in the direction of developing village and small-scale industries, in order to restore lost purchasing power and bring aggregate demand back at least to the levels of 1991, for the period 1991 to date has been the period of not only loss but a regression of rural livelihood. Finally, the restoration of employment and purchasing power has to be supported by sincerely implementing the employment guarantee programme.

NOTES

1. See Baker, Epstein and Pollin, 1998, Halevy and Fontaine, 1998, Patnaik, 2000.
2. The IMF and the World Bank are together referred to as the Bretton Woods Institutions or BWI in brief, since they were set up after the Bretton Woods Conference in 1944.
3. A rising contribution of services to GDP from an initial situation of a high share of industry to GDP has been typical for advanced economies. India however is seeing a fast shift to services from a relatively low initial share of manufacturing and mining output, less than 30 per cent of GDP, which is now down to about one-fifth. This shift to services reflects de-industrialization and worsening income distribution.
4. There are certain official norms of minimum stocks to be held, which vary according to the season.
5. Statement 9 on p. 21, Report no. 498 of NSSO on *Indebtedness of Farmer Households.*

REFERENCES

Baker., G. et al. (eds.) (1998), *Globalization and Progressive Economic Policy,* Cambridge: Cambridge University Press.

Bhalla, G.S. (2005), 'The Challenge of Food Security in India', D.S. Tyagi Memorial Lecture, 9 July.

Blyn, G. (1966), *Agricultural Trends in India, 1891–47,* Philadelphia: University of Pennsylvania.

Calvert, H.C. (1936), *The Wealth and Welfare of Punjab,* Lahore: Civil and Military Gazette.

Cornia, G.A. et al. (eds.) (1987), *Adjustment with a Human Face,* vol. 1, Oxford: Clarendon Press.

Darling, M.L. (1977), *The Punjab Peasant in Prosperity and Debt,* New Delhi: Manohar.

Government of India, Ministry of Finance, *Economic Survey,* 1995–6 to 2005–6.

Halevy, J. and J.M. Fontaine (eds.) (1998), *Restoring Demand in the World Economy,* Cheltenham: Edward Elgar.

Kalecki, M. (1972), 'Problems of Financing Economic Development in a Mixed Economy', *Selected Essays on the Growth of Socialist and Mixed Economies,* Cambridge: Cambridge University Press.

Karshenas, M. (1995), *Industrialization and Economic Surplus: A Comparative Study of Economic Development in Asia*, Oxford: Oxford University Press.

Kindleberger, C.P. (1987), *The World in Depression 1929–39*, Harmondsworth: Pelican Books.

National Sample Survey Organisation, Government of India.

Patnaik, P. (2000), 'The Humbug of Finance', Chintan Memorial Lecture, Chennai, India, 8 January. Available on the website (www.macroscan.org). Also included in P. Patnaik (2003), *The Retreat to Unfreedom*, Delhi: Tulika.

Patnaik, U. (2006), 'Poverty and Neo-Liberalism', 'Kale Memorial Lecture', 3 February, Pune: The Gokhale Institute of Politics and Economics.

______ (2005), 'Theorizing Food Security and Poverty in the age of Neo-liberal Reforms', *Social Scientist*, vol. 33, nos. 7–8, July–August.

______ (2004), 'The Republic of Hunger', *Social Scientist*, vol. 32, nos. 9–10, September–October.

______ (2003 c), 'Global Capitalism, Deflation and Agrarian Crisis in Developing Countries', *Social Policy and Development Programme*, paper no. 13, United Nations Research Institute for Social Development (UNRISD), October.

______ (2003b), 'Food Stocks and Hunger—Causes of Agrarian Distress', *Social Scientist*, vol. 31, nos. 7–8, July–August.

______, (2003a), 'On the Inverse Relation between Primary Exports and Domestic Food Absorption under Liberalized Trade Regimes', in J. Ghosh and C.P. Chandrasekhar (eds.), *Work and Welfare in the Age of Finance*, Delhi: Tulika.

______ (2002), 'Deflation and Deja-Vu', in Madhura Swaminathan and V.K. Ramchandran (eds.), *Agrarian Studies: Essays on Agrarian Relations in Less Developed Countries*, Delhi: Tulika.

______ (1996), 'Export-oriented Agriculture and Food Security in Developing Countries and India', *Economic and Political Weekly*, vol. 31, nos. 35–7, special number 1996. Reprinted in *The Long Transition: Essays on Political Economy*, Delhi: Tulika, 1999.

______ (1994), 'India's Agricultural Development in the Light of Historical Experience', in Terence J. Byres (ed.), *The State, Development Planning and Liberalisation in India*, New Delhi: Oxford University Press.

Paustian, P.U. (1930), *Canal Irrigation in the Punjab: An Economic Inquiry Relating to Certain Aspects of the Development of Canal Irrigation by the British in the Punjab*, New York: Cambridge University Press.

Reserve Bank of India (2006), *Handbook of Statistics on State Government Finances*, Mumbai: Reserve Bank of India.

Swaminathan, Madhura and V.K. Ramchandran (eds.) (2002), *Agrarian Studies: Essays on Agrarian Relations in Less Developed Countries*, Delhi: Tulika.

CHAPTER 5

Whom and What to Fight?: Indian Farmers' Collective Action under Liberalization and Globalization*

Staffan Lindberg

INTRODUCTION

The main character of the farmers' movements in India is that of a social movement, protesting and trying to influence agricultural policies by contentious actions.[1] Contemporary studies of such movements tend to stress three key dimensions in order to grasp their structure and dynamics:

> Increasingly one finds movements scholars from various countries and nominally representing different theoretical traditions emphasizing the importance of the same three broad sets of factors in analyzing the emergence and development of social movements/revolutions. These three factors are (i) the structure of political opportunities and constraints confronting the movement; (ii) the forms of organization (informal as well as formal), available to insurgents; and (iii) the collective process of interpretation, attribution, and social construction that mediate between opportunity and action. (McAdam et al., 1996: 2)

In these notes I assume that the farmers' movements in India have the character of a social movement and that, therefore, these

*This paper is written for the research project 'Farmers' Organizations and Agricultural Development in Taiwan, Thailand and India: The Challenge of Democratization and Globalization' undertaken together with Dr. Stig Toft Madsen, NIAS, Copenhagen. The project is funded by Sida-SAREC and Danida, and in India is affiliated to the Tata Institute of Social Sciences, Mumbai. The paper builds on fieldwork carried out in Punjab during March–May 1998, and February 2000.

three aspects are worthy of consideration. Of these aspects, political opportunities take on salient importance today in view of the shifting relationship between the state and the economy under globalization and a neo-liberal economic regime. But the 'new opportunities' may have important ramifications for both forms of organization and the 'cultural framing' needed for such action.

AN INDIAN VARIANT OF ADAM

After Independence India shared some central features of what could be called an Asian Development of Agriculture Model (ADAM). The basic features of this model are:

Everywhere smallholding farmers have, after more or less radical land reforms, been placed in a regional and national framework for increased production. This follows a pattern for agricultural development in which the state plays a dominant role.

I. Colonial systems of agriculture based on smallholder agriculture sometimes create the infrastructure and institutions for later development. Typical cases are Punjab developed by British colonialism and Taiwan developed by Japanese colonialism.
II. Post-colonial governments, supported by the US and international financial institutions, rely on heavy state intervention to increase agricultural production.

This pattern accounts for land reforms, irrigation systems, agricultural extension, research on and provision of inputs (seeds, subsidized fertilizers, and pesticides), supply of credit, marketing cooperatives, price regulations, the Public Distribution System with fair price shops, and so on. Farmers' organizations are developed by the state, in the form of farmers' associations and credit and service cooperatives emulated from the West.

Thus, ADAM is characterized by strong central state power, institutional reforms, and pervasive state intervention in and management of agricultural development. In an illuminating overview of these features, Djurfeldt and Jirström (2002: iii; Djurfeldt et al., 2004) conclude that the 'green revolutions in Asia were state-driven, market-mediated and farmer-managed processes'. They also point out that rather than merely being technical in nature, the Green Revolution was a political phenomenon in the context of the Cold War (Djurfeldt and Jirström, 2002: 1, 18–21).

In the case of India, a weaker variant of this model is discernible. The great ethnic and regional complexity and caste and tribal hierarchies make it hard to form and practice a uniform strategy everywhere. Equally important, India is, unlike most countries in Asia, a democracy dominated by strong interest groups like big capital, public employees, trade unions (formal sector), and a strong rich farm lobby, but also a democracy driven by increasing horizontal and populist mobilizations, as well as social movements of various kinds (women's, dalits', environmental). The result is a typical Indian compromise, which is weakened by poor management and sheer corruption, that is rent seeking (Djurfeldt and Jirström, 2002: 17; Harris-White, 2002).

Thus, the Indian variant has all the elements of ADAM, albeit operating with various degrees of effectiveness. Of all the features listed in the section on agricultural development above, land reform was evidently one of the weakest points, taking place more by default than as a result of strong popular mobilization and state intervention. As a result, the development of family farming took longer in gestation. A second feature, in sharp contrast to say Japan, is also the tendency 'to replace family farming with hired labour when the family gets the economic means to do so' (Djurfeldt and Jirström, 2002: 16). Thus, farms today are small, but to a large extent managed by gentlemen farmers employing hired labour.

It is important to note that India's relative progress in the development of agriculture is mainly confined to regions with assured irrigation, the Delta regions. Even the steady development of irrigation in the form of well irrigation is mainly confined to these regions. In the dry rain-fed regions, the so-called Deccan, there has been no corresponding development of new seed technology (e.g. for millets) or more advanced irrigation schemes. These regions, about three-fourths of the total area and home to about 50 per cent of the rural Indian population, have rather served as 'hinterland' supplying labour and resources (e.g. bullocks) to the Green

Revolution areas, developing further small-scale industry producing consumer goods for the growing urban and rural markets, and supplying some raw materials to industry. Thus, Deccan has assumed the role of agriculture as a whole in classical theories of industrial development.

At the peak of this model of development, around 1990, rural India was a divided continent. On the one hand were its Delta-based Green Revolution areas with fairly prosperous agriculture, fast growth of rural industry and services and with increasingly pluri-active farm households straddling between farming and non-farming activities. On the other hand, there were vast regions left behind, like Orissa, Madhya Pradesh and Assam with more than the average population below the poverty lines and widespread malnutrition especially among women and children (Radhakrishna et al., 2004).

FARMERS' MOVEMENT AND ADAM

With a strong tradition in peasant movements fighting for land to the tiller and fair rent and a broad base in the emerging and dominant middle and rich peasantry (after tenancy reforms), the new farmers' movement emerging in the 1970s began to contest government policies related to its guided marketization of agricultural development. With far-reaching state intervention relating to infrastructure, inputs, credit, extension and marketing of crops it was easy to mobilize farmers on various economic issues of vital importance to the individual farm economies. Typical issues were electricity tariffs for pump-irrigation, credit conditions in state institutions, price of wheat, rice and a few other staple crops regulated by government bodies.

Like in other parts of Asia, it was the post-colonial state that took the lead in the modernization of the economy. Emulating the past in the West, service cooperatives and farmers' associations were started that were controlled by the government. This is a major feature separating the Asian farmers' political context from that of farmers in the West, where these cooperatives were normally controlled by the farmers themselves. The Asian farmers, therefore, did not develop strong trade unions and cooperative interest organizations controlled from below. The farmers' movement took an explicitly political form agitating for better production conditions including prices of inputs, credit terms, tariffs and output prices. By and large they were non-party political movements, but because of their strength in the five key states (Tamil Nadu, Punjab, Karnataka, Maharashtra and UP) they managed to influence party politics and direct government policies in farmer-friendly directions (Varshney, 1995).

In the late 1980s the strong unity that had been there between the various regional chapters of the farmers' movement began to dissipate. The rising tide of neo-liberalism took hold on part of the farmers' movement itself. This development considerably weakened the movement when it had to adjust to a gradually changing economic situation and political regime.

IS GREEN REVOLUTION RUNNING OUT OF STEAM?

PUNJAB REACHING A PLATEAU

Being the most successful Green Revolution area in India, Punjab agriculture seems to have reached a plateau of 'relative stagnation' at a fairly high level of production. Wheat and paddy production is still dominating, with yields of wheat at between 15–20 quintals per acre, with an average application of about 150 kg of fertilizer per acre, and with farming highly mechanized. Bullocks are almost nowhere to be seen in ploughing or transport. Even relatively small farmers now have tractors and accessory implements, the cost of which today ranges between Rs. 2.5–3.5 lakh, a tractor being the norm for feeling like a real farmer. Combine harvesters are used for all harvesting of rice and most of the wheat. These are often locally made and operated by some enterprising farmer and/or artisan families (blacksmiths, carpenters, etc.). Small industries and workshops are everywhere in small towns and villages catering to the increasingly mechanized needs of farmers. The great number of tractors (*c.* 3,50,000), accessories, implements, and combine harvesters (4,700 units in 1998), many of them locally made, have stimulated the local industries and workshops all over the state. There are now so many combine harvesters that some are being sent outside the state to seek all year deployment.

According to the farmers and their leaders there has been very little improvement in plant breeding in the last decade.[2] Another reason for stagnation in crop yields may be poor rotation and lack of natural fertilizers with cow dung being used for other purposes, mainly as fuel. On the other hand, cost of production increases with increasing mechanization and higher input costs. Pesticides, insecticides, herbicides, and weedicides, are being applied with very little state regulation and control, making for high costs and threatening environmental damage. Prices of staple crops, on the other hand, have stagnated according to the farmers interviewed, who say that the 'profitability' of farming these crops is low.

Again, according to the farmers and Bharatiya Kisan Union (BKU) leaders, increased cost of production, mechanization of family farms, and stagnating prices of their produce have landed farmers in a debt trap. A survey of the Institute of Communications and Development in Punjab in October 1997 showed that farmers had a total debt of more than Rs. 5,700 crore (*The Hindu*, 1 March 1998).[3] The major part of these loans, or 62 per cent, have been taken from informal sources such as commission agents (*arhtiyas*), traders and moneylenders with interest rates ranging from 24–48 per cent.[4]

Institutional credit with its lower interest rates now plays only a minor role in the overall financing of the farm economy. Thus, it seems that state intervention and supply of credit through co-operatives and rural banks have not lead to a retreat of usurious capital, rather in its latest stage the Green Revolution and mechanization seem to have given it a boost. For long-term investments the farmers depend on institutional credit, while for cultivation costs and social expenses they make frequent use of loans particularly from the commission agents. Rising aspirations of the farmers in consumption and social status (higher dowry, and so on) are part of the reason for their high indebtedness.

Another significant development is the changing work culture of the Punjabi farmers. The development of wheat and rice cultivation has brought higher incomes to almost all sections of the farmers except, perhaps, the very small and marginal farmers. This has changed their pattern of participation in agricultural work. In the pre-Green Revolution era the farmers participated manually in almost all operations. The thrifty farmer used the agricultural implements himself lest the worker may spoil them. With prosperity coming in as a result of capital-intensive agriculture, the farmers started withdrawing from active operations and most of them have increasingly become farm managers, administering and supervising the work done by the hired wage labourers from UP and Bihar. The proud Sikh farmer on his tractor pulling the trolley of wheat to the *mandi* used to be a common sight earlier. But that is not a typical sight anymore. His withdrawal from production activities no doubt gives him more free time to involve in agriculture supportive and administrative work (buying inputs, dealing with authorities and credit institutions, marketing) and in political activities too, such as those connected with the farmers' union (and, in the case of the young, in militant actions!). It also means an additional financial burden on the farm in terms of higher labour and machine costs. The size of a commercially viable farm has thus moved upwards.

The small and marginal farmers may have gained in productivity with intensification of agriculture through hybrid seeds but have lost in viability with mechanization of production, which has brought a new economy of scale into operation. Joint farming among brothers sharing a tractor and other implements is one solution to the increased capital inputs into modern agriculture. It is becoming, otherwise, not only unmanageable but economically unproductive. Consequently, the small farmers are increasingly renting out their land to other farmers who need to complement their landholdings with more land to make full use of machinery and labour. The small farmers still hold on, however, either as milk producers or as wage workers, often outside the village, since for a Jat Sikh farmer to work for a better endowed caste mate in the same village is not considered honourable.

Liberalization

In 1991, the Indian polity turned full circle on its previous strong policy of import substitution and self-reliance. 'Fast track liberalization' was in the air. For agriculture, this would mean

> To phase out in ¾ years, the controls over agriculture sector to enable markets to function efficiently and to eliminate the economic distortions caused for reasons of private interests by bureaucrats and politicians. At least 20 major agricultural commodities would be involved, notable food grains (whose domestic price would rise to the level of the world markets), together with oilseeds and sugar should the OECD countries fulfil their treaty obligations under the GATT and reduce their agricultural protection. (Harriss-White, 2002: 7, quoting Randhawa, 1994: 353–78)

It would also mean, again according to Harriss-White:

> For India this was an agenda for the deregulation of commodity and factor markets. It involved the elimination of movements restrictions, the unbiased operation of freight transport, the privatisation of storage, the deregulation of agro-processing from its special (protected or restricted) status as a 'small scale industry', the dismantling of subsidies on fertiliser and electricity, with increased exports compensating for the production disincentives resulting from the price squeeze from raised costs of production. It involved either dismantling or privatising most of the activities of the public distribution system, the liberalisation of the land market—starting with land lease deregulation—and permission for corporate investment in 'wasteland' and degraded forest. (ibid.: 7)

In reality, while there has been a slow but visible liberalization of the conditions for industries and services (cf. Degnbol-Martinussen, 2001), there have been fairly little such changes in agriculture.

This can readily be seen from the 2003 World Bank report on the Indian economy. During the 1990s, subsidies to the agricultural sector (fertilizer, electricity, irrigation, credit and food) were growing, despite attempts to downsize them all. In 1997 the Public Distribution System (PDS) was revamped in the Targeted PDS which seems to have 'increased participation of persons below the poverty line', but recently there has been a proposal to bring back the untargeted PDS again, which according to the World

Bank 'is likely to bring back the earlier problems of the PDS of subsidies being captured by non-poor households and will likely result in an escalation of food subsidies' (World Bank, 2003: 68).

Likewise, there has been a steady increase in the minimum support price (MSP) for wheat and rice despite attempts to keep it down. The political pressure from the states where most of the procurement takes place (like Haryana, Punjab and Andhra Pradesh) has simply been too strong. This has resulted in growing buffer stocks—which stood at 60 million tonnes in July 2002—downward pressure on open market prices, and 'increased use of food grains for welfare schemes and drought relief . . . and subsidized exports' (ibid.: 67).

The fertilizer subsidy has also been maintained with its unbalanced emphasis on subsidies for urea with negative effects on soils. Recently, however, decisions have been taken to deregulate the fertilizer industry and phase out subsidies. It remains to be seen if this will be fully implemented.

Most states supply electricity to farmers without any charges at all. This has resulted in a financial crisis for the state electricity boards, which have no funds for investments or even maintenance. So power supply is erratic, which has the effect that farmers are even less willing to pay for it.

The growing bill for all these subsidies has, according to the World Bank, 'crowded out' other types of state intervention in the agricultural sector. The most growing is public investment, 'its share of annual agricultural capital formation decreasing from 44 per cent in 1985–6 to 23 per cent in 2000/1' (ibid.: 66). This has especially affected the construction of surface irrigation systems and roads, despite ambitious plans in these areas. Other important public interventions like research and extension services, as well as credit to farmers have also been affected by the growing burden of subsidies.

Harriss-White studied agricultural policy and practice in Tamil Nadu during the 1990s and found that

> the state's involvement in agriculture was expanding, not contracting, in the early phase of liberalisation. (Harriss-White, 2002: 11)

Above all, the state agenda expresses continuity, complexity and expansion. It reflects solid bureaucratic interests, a nexus structured around patronage and the application of scientific techniques, together with the needs of a diverse but powerful political constituency. (ibid.: 10)

Tamil Nadu's stated policy for agriculture is broad and all-encompassing. Moreover it involves almost all government departments one way or the other, and a vast fora of international aid agencies, and local and international NGOs each pursuing their pet projects. There is little coordination and the state's capacity to absorb these project funds is saturated with incomplete uptake and delays. All this results in an unfocused development policy for agriculture, 'the garden' to use Harriss-White expression, which seems to have lost its major objectives:

Major parts of the agricultural economy are strikingly marginalised on this agenda. The food grains economy of rice and groundnuts, in which a state with fabled grain bowls has lost rank and where growth has faltered, are not the central concerns of local agricultural policy. (ibid.: 11)

Harriss-White in analysing why there is this inertia comments:

This is the intrinsic complexity of agricultural policy, the range of interests in it being multiply justified, inconsistently layered, path-dependent and contested, operating simultaneously in policy paradigms, involving most departments of government, linking the jurisdiction of centre and state. Laws and office procedures mirror this complexity and the relevant allocations of skilled man-power are also extensive. Organising the wholesale dismantling of existing agricultural policy has simply not been imagined. The policy process may be much more influential in blocking reforms than the sheer size of subsidies. It is also the expression of the unusual common interest of competing class forces. (ibid.: 12–13)

If this is the situation in Tamil Nadu, what then happens in Punjab, one of the core regions of the Green Revolution? From a recent interview with Sucha Singh Gill (2004) it appears as if the state administration has lost interest in agriculture. And whatever focus there is, it is not being implemented properly. According to Gill:

> In fact, in the absence of scientific crop-mix corresponding to the ground or surface water, much could not be expected. The Punjab Agricultural University is supposed to be flush with funds to undertake research in bio-technology for crop productivity improvement. This is not happening. The result is low yields and un-remunerative returns, particularly to the small or semi-medium and medium farmers, who form the bulk of 12 lakh farming families . . . the 10th Plan document has suggested that agricultural research and extension be entrusted to the private or the corporate sector. In practice, the ordinary dealers and *arhtiyas* have become nodal agents on behalf of the absentee private-corporate companies even in 'contract farming'. These dealers, in the name of 'contract farming', supplied spurious seeds to the farmers at a high cost and offered low procurement price for whatever little was produced. The farmers had burnt their fingers. Though the corporate sector had been entrusted with this responsibility, it did not have requisite basic infrastructure, like seed-testing laboratories. Otherwise, how come the winter maize did not grow? (Gill, 2004)

What then happened to the 'successful' Green Revolution of the 1970s and 1980s? First of all, the policy of Green Revolution seems to have reached a kind of threshold in terms of environmental constraints. Radhakrishna in a perceptive analysis lists such factors as being: (i) deceleration in annual growth rate of area under high-yielding varieties, (ii) falling soil fertility due to intensive mono-cropping, and (iii) over-exploitation of ground water due to the unregulated expansion of tube wells (Radha Krishna 2002: 245). The first of these factors is mainly due to lack of new suitable areas for the cultivation of high-yielding wheat and rice, which is partly due to the slower expansion of irrigation facilities. It appears that the Delta cultivation has reached its outer limit.

At the same time consumer demand for food is gradually shifting in the direction of non-cereal items. Vegetables, fruits, milk and meat make up a fast increasing share of the food basket, while those underfed are not able to demand even the bare necessity of grain. The demand for cereals has stagnated with a downward pressure on output prices (Patnaik, 2004). In this situation farmers have started diversifying into other crops, some of these cultivated on contract for agro-business corporations of various kinds. These markets are characterized by unstable prices. To this should be added the opening up of international trade in agricultural commodities along with less effective state intervention in the markets for farm inputs and produce leading to strongly fluctuating prices.

Thus, while ADAM peaked in India in the 1980s, the 1990s has seen a gradual weakening of the strategy, as evidenced in particular by decreasing public investment in agriculture (infrastructure, research). The motivation to continue with ADAM has become weak. The increasingly diversified agricultural economy, with a multiplicity of crops and other products, cannot easily be managed in a 'command mode' like that of ADAM. It was the 'simplicity' of the agricultural economy with focus on wheat and rice that allowed for an overall simple state intervention and management strategy.

DIVERSIFICATION AND THE NEED FOR A 'SECOND GREEN REVOLUTION'

With the fast diversification of agriculture into a variety of crops and products, a multitude of actors, national as well as international, state as well as corporate, and a complex of markets have emerged on the scene. One indication is the development of new seeds by the domestic and international companies and agencies including the use of biotechnology in plant breeding despite widespread protests and attempts to ban it altogether. None of these processes are managed or even orchestrated by the state. Instead, the insufficiency of current state intervention is glaring.

However, a balanced 'second Green Revolution' addressing the development needs in the Deccan regions, with new types of irrigation systems, and with new seeds of a variety of crops, vegetables, fruits, etc., cannot be state driven in as comprehensive a way as the first Green Revolution.

It is here that we can see most clearly that the Indian agriculture stand at a crossroad. The ADAM has peaked and is being slowly dismantled. What will replace it is far from settled? From a political point of view there are many issues to be discussed and clarified by the farmers' movement.

THE POLITICAL SITUATION

With the decentralization of the political system, power and resources are meant to be devolved to lower levels of administration, ultimately to the village panchayat level. However, it is somewhat of an irony that this opening up for local participation takes place in a situation when the state, at least partly, withdraws itself as an active agent in the processes of development and social welfare. Fewer and fewer resources are given, and those that reach, are often handled by an NGO, which normally lacks accountability to the public it is supposed to serve. Much of the traditional politics is emptied of meaning under neo-liberalism, '. . . where we see diminution of national and local governing structures, erosion of established ideologies, de-politicization of social movements and NGOs, . . . ' (Boggs, 2001: 3). Is this political 'atrophy' inevitable?

Another factor with a strong bearing on the farmers' movement is the fact that many of its members are not full-time farmers anymore. It may be that the farming community, inclusive of the landless labourers, still constitute about 60 per cent of the population in India. Agriculture, however, accounts for less than 25 per cent of the national income. But these figures hide an increasingly important phenomenon: the integration of agricultural population into other sectors of the economy, that is, industry and services. Today, most of the agricultural households are pluri-active, straddling between farming and other activities, whether as seasonal labourers or small-scale entrepreneurs in the local economy, or with sons and daughters working as labourers in more or less distant locations. Moreover, about one-third of all rural households do not get their entire income from industrial and service work (Goswami, 2004).

Agriculture and farming are no more an all-encompassing way of life and identity. The politics based in such an identity does not have the same meaning as it had for the peasant movements against the feudal and colonial orders. The non-agricultural, urban, metropolitan, global views and values today mix with inherited ones. It is a period of rapid change, involving even uprooting of people.

Their heightened anxiety is an order of the day. And the ideological reaction created is not as much as a simple going back, but as an invented religion and morality. But farmers' politics cannot emulate this if it wants to look forward to and defend the position of agriculture and its people.

In this situation what are the 'political opportunities' for farmers to act upon, if any?

QUESTIONS RATHER THAN ANSWERS

A number of issues related to the development of agriculture in the post-Green Revolution phase need to be addressed:

1. *What should be the role of the state, given India's increasing integration with the world economy and the fast diversification of agriculture and agro-processing industries?*

The starting point here is the assumption that the general trend of integration of the Indian economy is going to proceed even if with considerable opposition. A reversal is not in sight. However, realistic alternatives are not being formulated taking into consideration the level of integration already at hand. India today wants to be a modern economy with access to the latest technology due to its increasingly advanced scientific development in some fields, and she needs to make use of the most sophisticated technology to solve food-security problem as well as the serious environmental issues facing the subcontinent.

Indian farmers need to make a priority list of the desired governmental interventions, also taking into consideration their financial implications. Agricultural income tax and fees for public services like supply of electricity and irrigation must also be discussed in this context since the scope for assistance from other sectors of the economy and foreign aid are dwindling with the growth of industry, information technology and other services as the core sectors of development.

What should, for example, come first, fertilizer subsidies and high and stable prices on farm produce or infrastructure, seed development, extension and credit? Ideally, of course, all of these simultaneously.

Recent history suggests that successive governments actually have been trying to do all this even if the political will has been weaker than during the hey-days of the Green Revolution. The political demand for subsidies and high support prices has however been much stronger than the demand for infrastructural investment. As we saw above, the World Bank experts also suggest that the high level of the former have 'crowded out' the latter.

But recently crops have failed and prices have crashed despite continued attempts at price intervention and market controls. Liberalization and the transition to a second Green Revolution away from the simple mono-cropping of the Green Revolution points to a dilemma here. The government no longer possesses the political and macroeconomic capacity to control and protect the agricultural economy when it is being increasingly diversified and driven by a great variety of domestic and international actors.

2. *Who can protect the farmers, if not the government?*

If the government cannot protect the farmers, then the farmers must protect themselves.

The most obvious method is to form crop-specific unions or growers' associations that can deal with the opponents in the market, viz., traders, contract firms, exporters, state agencies, and other buyers of the agricultural produce. This has been the way of the farmers in all the industrializing countries of the world. In East and South-East Asia there is already such a development under way, although greatly hampered by the presence of traders and agribusiness interests in the leadership of the unions (cf. Lindberg, 2004). A separation of the farmers' and traders' representation is absolutely necessary in order to avoid a corporatist system with its hiding of dominant interests, which in reality weaken the bargaining position of the producers.

How far has this type of organization reached in Punjab?

Closely coupled with organizing the 'growers' associations' is the capacity to make one's own studies of specific economic and other issues in order to negotiate successfully. Such a research unit could also assist individual farmers in relation to the traders and dealers with, for example, legal advice. It could also generate a distress fund to aid the farmers facing economic hardships for various reasons. An organizational form for such an activity could be a member-based NGO that could also receive funds from other sources.

3. *But what should the government do if it cannot regulate and control prices fully?*

The main role of a government is to ensure the extended reproduction of overall conditions of agricultural production. This means undertaking planning and investment in land use, irrigations, roads, power supply, and so on.[5] In these areas of state management there is now a reform process (for irrigation see, for example, Gulati et al., 2005), that farmers' unions need to actively relate to.

Institutional reforms also belong under this category, foremost of them perhaps land reforms and land consolidation. These reforms are politically driven and depend on the strength of the driving forces and their alliances in the rest of the society. Currently, it appears that moves to allow corporate farming are much stronger than interest in further land reforms, at least in the Delta areas.

Under the Green Revolution the state made strong inroads into credit supply and the production and sale of seeds, fertilizers and pesticides. Much of this was organized through service cooperatives controlled by the state. These were always criticized for being inefficient and corrupt, but there is a high probability that historical evaluations will recognize their strategic role in the making of the Green Revolution.

Of these, credit is probably the strongest case for continued state intervention today, not least due to the strong role of the state in the banking system. Here the state needs to retake the initiative and massively fund farm operations through its various lending institutions. This is the only way to tame usurious capital. It is also hard to imagine that the large number of NGOs with micro-credit schemes could make up for the vast and generalized needs of farmers in this respect.

When looking at other inputs, private and corporate actors have taken over most of the initiative. Here the role of the government is moving in the direction of regulation and control. An exception to this could be if strong public institutions, universities and seed farms could retain and even further develop their capacities to compete with private firms, domestic as well as international. In combination with a strong monitoring role, such an institutional arrangement may ensure a more balanced development.

Government regulation and control is also the policy that may be appropriate for the increasing use of biotechnology in seed development. At least farmers need to take a firm stand on this issue.

The use of biotechnology is going to be one of the main characteristics of the second Green Revolution globally. Most researchers and seed developers see this technology as a powerful and quick method to develop new seeds with superior qualities, like pest- and drought-resistance and with added nutritional values. It is not regarded as a qualitatively new technology compared to earlier methods of plant breeding.

Ideally, there may be a strong case for banning most of it until it has proved its credentials, that is, until prolonged testing has proved that it is neither harmful to humankind nor to the environment. However, the speed with which the new technology is developing in the world today, and even in Europe (where most people still are very critical) makes it impossible to ban its introduction in countries like India with open borders and erratic state policing. Recent history illustrates the point. Rather than living with an illegal under-vegetation of uncontrolled seeds imported or produced inside India and used by farmers who see them as superior to what is offered in the 'legal' market (which has caused quite a lot of harm recently), the farmers' movement needs to discuss the option of strengthened state control and regulation of these new seeds in order to ensure as balanced a development as possible under Indian conditions. The alternative may play havoc with the farmers and most probably, prove to be a setback to Indian agriculture in terms of competitiveness with the rest of the world.[6]

4. *Could the government do anything about the distribution of income so as to ensure that the poor can also afford to buy food?*

Food-for-work programme geared at the physical infrastructure for agriculture is probably the most effective way of combining the two aims, namely the strengthening of the agricultural production capacity and the creation of an effective demand among the poor, thus stabilizing food prices. In the current discussion it is suggested that existing subsidies on the PDS, fertilizers and liquified petroleum gas (LPG) should be dismantled in favour of massive food-for-work programmes. Could the farmers accept such a drastic policy change?

5. *Role of the cooperatives*

In the West cooperatives are organized by the farmers themselves. In India there is a rather different story as the state controls most of them. Today there are legal opportunities to free cooperatives from this straightjacket. How feasible is it for farmers themselves to run cooperatives for supply of various inputs and milk production? Can MARKFED and other cooperatives at all compete with other more or less powerful actors on these markets?

6. *Role of the NGOs*

The growth of NGOs has lead to the fragmentation of what used to be the generalized state administration and services and to a weakening of claims for rights due to the agenda setting of various donors. In Mary Kaldor's terminology, the NGOs are 'tamed' social movements (Kaldor, 2003). They are meant to solve the problem of erratic and inefficient bureaucratic action but at the same time, it is evident that they greatly disturb political representation. However, whatever their political and other effects, they seem to have come to stay. Farmers' and other social movements will have to deal with this reality.

7. *Political representation and the driving force in economic organizing?*

Is a politically united farmers' movement necessary for the creation and/or strengthening of a 'family' of farmer-driven economic organizations required to put pressure on the government and ensure legislation?

The dilemma today, as often, is that other political interests have come to play an important role in the political organizations of the farmers. This has lead to a split, which weakens the overall power of the movement. But from where should the farmers' movement actually get its ideological orientation? Is there an ideological frame, which could fit them all?

INSTEAD OF A CONCLUSION

Will there eventually be a new agricultural policy capable of addressing all those aspects of sustainable agricultural economic development, which as everybody is aware of, cannot be managed by individual farmers or by the corporate sector? A policy that is able to direct state intervention to make laws, enforce laws, invest in general infrastructure like irrigation, power, roads, etc.? A policy which takes a critical stance on globalization, capable of handling complex issues of trade, biotechnology and intellectual property rights in seeds in modes conducive to the particular conditions of Indian agriculture and its people?

Part of the answer may lie in adequate collective action by the farmers themselves.

NOTES

1. Della Porta and Diani define a social movement in the following way: 'We will consider social movements—and, in particular, their political component—as (1) informal networks based (2) on shared beliefs and solidarity, which mobilize about (3) conflictual issues, through (4) the frequent use of various forms of protest' (1999: 16).
2. Most of these interviews were carried out in 1998 and 2000.
3. '. . . if the principal amount of the entire debt was to be liquidated, it would result in the sale of 3.31 per cent of the total area owned by the farmers in the state, at the current average per acre price of land.' (ibid.) This indebtedness may not seem high by Western standards and considering the high value of land in Punjab, but it seems to be high by Indian standards.

4. An alternative estimate puts borrowed amount from private sources at 46 per cent (*The Hindu*, 1 March 1998).
5. One can even argue that education and health investments are of crucial importance for agricultural and rural development.
6. For an informed discussion see Herring, 2005 and Cohen, 2005.

REFERENCES

Boggs, Carl (2001), 'Economic Globalization and Political Atrophy', *Democracy & Nature*, vol. 7, no. 2, pp. 303–16.

Cohen, Joel I. (2005), 'Poorer Nations Turn to Publicly Developed GM Crops', *Nature Biotechnology*, vol. 23, no. 1, pp. 27–33.

Della Porta, Donatella and Mario Diani (1999), *Social Movements: An Introduction*, Oxford: Blackwell Publishers.

Degnbol-Martinussen, John (2001), *Policies, Institutions and Industrial Development: Coping with Liberalisation and International Competition in India*, New Delhi: Sage Publications.

Djurfeldt, Göran and Magnus Jirström (2002), 'Asian Models of Agricultural Development and Their Relevance to Africa', Working Paper, Lund: Department of Sociology, Lund University, February.

Djurfeldt, Göran, et al. (eds.) (2004), *The African Food Crisis: Lessons from the Asian Green Revolution*, Wallingford, England and Cambridge, MA: CABI Publishers.

Editorial (2002), 'Agricultural Policy: Need for Paradigm Shift', *Economic and Political Weekly*, 9 March.

Gill, Sucha Singh (2004), 'Punjab Facing Deep Crisis', Interview in http://www.onlypunjab.com/latest/fullstory-newsID-1896.html

Goswami, K. (2004), *Changing Contours of Rural India*, CERG Advisory, http://www.worldisgreen.com/2004/09/15/changing-contours-of-rural-india

Gulati, Ashok, Ruth Meinzen-Dick and K.V. Raju (2005), *Institutional Reforms in Indian Irrigation*, New Delhi: Sage Publications Inc.

Harriss-White, Barbara (2002), *Development, Policy and Agriculture in India in the 1990s*, Oxford: Queen Elizabeth House Working Paper Series No. 78.

Herring, Ronald J. (2005), 'Miracle Seeds, Suicide Seeds, and the Poor: GMOs, NGOs, Farmers, and the State', Raka Ray and Mary Fainsod Katzenstein (eds.), *Social Movements in India: Poverty, Power, and Politics*, Lanham, Boulder: Rowman and Littlefield Publishers Inc., pp. 203–32.

Kaldor, Mary (2003), *Global Civil Society: An Answer to War*, Oxford: Polity Press in association with Blackwell Publishers.

Lindberg, Staffan (2004), 'When Will the Farmers Unite? The Long Road to Democracy in Rural Thailand', *The Indian Journal of Labour Economics*, 47(3): 525–39.

McAdam, Dough, et al. (1996), *Comparative Perspectives on Social Movements: Political Opportunities, Mobilizing Structures, and Cultural Framings*, Cambridge: Cambridge University Press.

Patnaik, Utsa (2004), 'Rural India in Ruins', *Frontline*, vol. 21, no. 5.

Radhakrishna, R. (2002), 'Agricultural Growth, Employment and Poverty: A Policy Perspective', *Economic and Political Weekly*, 19 January.

Radhakrishna, R. et al. (2004), 'Chronic Poverty and Malnutrition in 1990s', *Economic and Political Weekly*, 10 June, pp. 3121–30.

Randhawa, N.S. (1994), 'Liberalisation and Implications for Agricultural Policy', in G.S. Bhalla (ed.), *Economic Liberalisation and Indian Agriculture*, New Delhi: Institute for Studies in Industrial Development, pp. 353–78.

Varshney, Ashutosh (1995), *Democracy, Development and the Countryside: Urban-Rural Struggles in India*, Cambridge: Cambridge University Press.

World Bank (2003), *India Sustaining Reform, Reducing Poverty*, A World Bank Development Policy Review, Poverty Reduction and Economic Management Sector Unit, South Asia Region.

CHAPTER 6

Farming Crisis and Farmers' Suicides in Punjab: An Examination of Institutional Dimensions

Sukhpal Singh

INTRODUCTION

The primary sector (mainly agriculture and livestock) accounts for 35.95 per cent of the Gross State Domestic Product (GSDP) at current prices in Punjab (*SAP*, 2006). In 2002–3, food grains comprising cereals and pulses accounted for almost 80 per cent of the gross cropped area (GCA) of the state. The only other crop, which accounted for a significant percentage of the GCA was cotton (6.3 per cent). By the late 1990s, 96 per cent of the gross sown area was irrigated, and 100 per cent of the wheat area and 94 per cent of the rice area were under high yielding varieties (HYVs). The state had more than 11 lakh tube wells and about 4.5 lakh tractors by 2004–5 (*SAP*, 2006). Further, 87 per cent of the paddy area and 52 per cent of the wheat area in the state were being harvested by combine harvesters. More than 80 per cent of the paddy straw and about 50 per cent of the wheat straw was being burnt by the farmers (Sidhu et al., 1998). There is overcapitalization in the agricultural sector of the state. The capital assets are grossly underutilized due to low and decreasing size of the farms. For example, in the case of tractors, the average annual use is only 400 hours compared with the minimum viable use of 1,000 hours. Due to this, the fixed cost component in crop cultivation cost has increased. This has hit hard the profitability of the farm enterprise, especially of small and marginal farmers who over-invested in farm equipment (Gandhi, 1997). From the market side, these high costs have led to the wheat and rice production in Punjab becoming either only marginally competitive or non-competitive in the international market (Gill and Brar, 1996). In fact, wheat cultivation in Punjab had become unviable in the mid-1980s itself when a farmer in Madhya Pradesh was able to earn higher net profits by spending less on modern inputs than a farmer in Punjab (Singh, 2000).

Water has been used highly inefficiently in Punjab agriculture. In the mid-1980s, there was wastage of water to the extent of 30 per cent (Bhatia, 1987). In Punjab, water table is falling by up to one metre per year (Lester R. Brown, 2003, www.earth-policy.org). Already 90 per cent of the 138 blocks of the state have been declared black in terms of water table falling at an alarming rate in these areas (Sidhu, 2002). It is found that 93 per cent of the paddy farmers had tube wells and 44 per cent of them had been affected adversely due to the decline in water table. These farmers had to sometimes deepen their tube wells or place the pump sets at a lower place in the well itself. This meant an extra cost of Rs. 2,000 for deepening the tube well during the early 1990s and Rs. 4,000 during the late 1990s compared to that during the 1980s (Singh and Kalra, 2002). Now, there is excessive dependence on ground water for irrigation which accounts for 70 per cent of the irrigated area.

The policy of free power supply to tube wells seems to have exacerbated the problem between 1997 and 2002. The zero pricing of canal water was a disincentive to farmers to conserve water. The south-western districts face problems of waterlogging and salinity, and soil nutrient deficiency created by excessive use of water. Cotton was, thus, replaced by rice. A recent study of 100 farmers of different categories (both diesel pump owners and electric pump owners) in two districts of Punjab (Ludhiana and Bathinda) revealed that for nine months, farmers pay higher electricity charges because of the flat rate system of payment than they would have to pay under the meter system. Farmers of all types were giving as many as 27–30 irrigations to paddy and five and six irrigations to wheat and cotton respectively. But water productivity in terms of produce per acre per cubic meter of water was lower in Punjab compared to that in UP for both paddy and wheat (Batra and Singh, 2003). Also, the overextension of rice cultivation led to increase in insect and pest attack with 57 per cent farmers reporting this problem (Singh and Kalra, 2002).

The other indicator of crisis in the farm sector in the 1990s has been the emergence of second-hand tractor markets where farmers sell old and new tractors to meet other requirements of funds. There are more than a dozen daily, weekly, and fortnightly markets for the sale and purchase of second-hand as well as new tractors in various market towns in the Malwa (southern Punjab) region and more specifically in the cotton belt of the state where thousands of farmers participate in selling or buying tractors (Singh, 1999; Rangi and Sidhu, 2000). In a state which today has more than 4.5 lakh tractors, accounting for 13 per cent of the total number of tractors in the country (13.62 lakh) in 2002–3 (*SAP*, 2006), with just 2.5 per cent of cultivated area, this phenomenon is disturbing. With only about 11.17 lakh operational holdings in the state, it means that every third holding in the state is equipped with a tractor. This is a situation of overcapitalization of the farm sector when more than 70 per cent of the farms are below 10 acres each. At the structural level, these problems have led to reverse tenancy, lack of effective farm industry linkages and suicides by farmers.

Even at the aggregate economy level, the growth rates of Net State Domestic Product (NSDP) from agriculture, total NSDP and the per capita income during the 1990s have been lower than that for the country as a whole (Chand, 1999). The growth rate of NSDP was only 2.72 per cent per annum during this period compared to 3.42 per cent during the 1980s and one of the lowest among Indian states, and lower than the national average of 4.02 per cent. In fact, SDP from agriculture grew at just 0.37 per cent during the 1990s and the whole of the primary sector growth was accounted for by the livestock sector which grew at more than 5 per cent per annum during this period. Though the state is next only to Kerala in overall Human Development Index (HDI), it ranks seventh in literacy. The development expenditure in the state as proportion of NSDP has stagnated in the 1990s (Singh and Singh, 2002).

A part of the problem of the state's farm sector can be traced to the agro-processing sector. A study of the primary food processing units (paddy milling and oilseeds milling) in the early 1990s revealed that neither were they dependent on the local economy for labour nor for investible surpluses in a direct way (Singh, 1999a). As regards raw materials, the industry was more a consequence than a cause of the cropping pattern in the state.

Thus, the main problems of the farm sector include declining growth rate of farm production, declining capacity of the agricultural sector to absorb labour because the employment elasticity with respect to output in agriculture has come down to 0.2 per cent, monoculture of wheat and paddy which account for more than 76 per cent of the GCA of the state, and overcapitalization of the farm sector. But, agriculture is the primary engine of growth without which Punjab will neither be able to accelerate growth nor achieve fiscal sustainability. As is clear from the above discussion, the economic condition of a vast majority of farmers, especially marginal and small, cannot be improved unless there are changes in the cropping pattern and the technology of production. Diversification within agriculture is intended to stabilize incomes and employment in the farming sector through cost reduction either directly through more rational input use or by higher yields or value realization improvement. The diversification can either be in terms of a variety of crops grown or technologies used for the same set of crops. Besides, it calls for certain institutional solutions to the problems of Punjab's rural and agricultural development. The first section of this paper examines the question of institutional dimensions of development with special reference to Punjab and as a case of institutional innovations. The next section discusses the role of institutions in development from a theoretical perspective. The cases of institutional innovations in India and in Punjab are reviewed in the third section. The last section concludes the paper with a discussion of possible institutional innovations required in the state's farm and rural sector.

ROLE OF INSTITUTIONS IN DEVELOPMENT

The role of institutions in development has been well recognized in the literature. Institutions give shape to the resources and their use. Indian agriculture has traditionally been dominated by statal and parastatal institutions and organizations which are more known for their failures than successes, though they did play an important role in the initial stages of growth of agriculture. Institutions or organizations play their developmental role by way of mobilization of local resources and/or their more efficient use for concerns that are locally relevant. In the emerging environment of reduced role of the state in agricultural marketing and input supply, due to reduction of subsidies and increasing space being provided to the private sector under liberalization and globalization policies, it becomes imperative to explore the role of the institutions and organizations in delivering better value to the farmers. In this context, it is worthwhile to look at non-state sponsored or encouraged institutional experiments, as they tend to be more locally relevant and sustainable.

However, with the onset of an environment of deregulation and market integration, the agri-business sector in India, more so in Punjab, needs more market-oriented, purpose-specific and professional organizations and institutions. Some of these kinds of institutions are already emerging on the scene in various sectors like irrigation, processing and marketing. The new institutional economics had once again brought to the fore the role of institutions in development. Institutions along with resource endowments and technology are three major factors in economic development. Institutions include social, cultural, economic, political and legal mechanisms which enable, impede or disable individuals, enterprises or communities in what they can do or do to improve their well-being (Baumol, 1990; Shah, 1995).

Emphasizing the need for institutional or organizational improvements for agricultural development, Tarlok Singh writes:

> for all but a small proportion of farmers and some favoured regions, the prevailing organisations are not in tune with the technological, economic, and human imperatives of modern scientific agriculture which involves continuing

improvements in technology, increasing use of industrial inputs, rising levels of skills and knowledge, and greater social control over scientific management of land, water and energy resources. (Shah and Vakil, 1979)

This need has only heightened in the present decade. Alagh (1982) is also emphatic about the role of institutions when stating:

We really have very little insight into the reasons which lead to the growth of some regions and stagnation in others. Production function approach gives desirable outcome but not the reasons that trigger them off. Very often perverse institutional conditions are described as constraints. To begin with, there is of course the purely factual question needing study namely, has agricultural growth taken place under different institutional conditions or not. It is, of course, true that some major agricultural problems being faced today, e.g. more efficient water use or application of new technology in dry regions, lead to answers which lie in the institutional plane. (Alagh 1982: 423)

Even later, he places the question of organization of agricultural development at the centrestage of agricultural planning since organizations and institutions alone make planning a feasible and operational exercise (Alagh, 1990).

Regional culture, history and social organization affect agricultural development by focusing on the amount and kind of value placed on material advancement, development skills and enterprise and allocation of property rights (Shah, 1995). That cultural endowments are at the root of institutional innovations is underlined by the fact that in regions known for enterprise and innovation, the cultural traits are found to be different from those in other regions. A recent study comparing Gujarat and Orissa for cultural influences on pro-economic attitudes found that Gujarat had much higher levels of effort and attitude to risk, and much lower levels of fatalism and need gratification (Biswas and Biswas, 1997).

Various forms of markets and the participants therein are important institutions for agricultural development. The markets are institutional arrangements that facilitate exchange of goods and services, thus allowing division of labour. In the upcoming liberalized economic environment for the agricultural sector, it is important to understand the efficacy and efficiency of existing market institutions as well as the need for new institutions to move from traditional- and domestically-oriented systems of production and marketing to modern and globally-oriented systems. In more specific terms, these aspects pertain to the procurement, processing, and marketing of farm products and agricultural inputs. Some of the arrangements in the output markets include contract farming, and producers' organizations for processing and marketing of agricultural produce and state and private arrangements for input provision and transactions in these markets.

However, at the more micro and specific crop or input level, there have been many developments in many areas of the country, which are in the nature of institutional reforms or innovations. These innovations need to be understood for their effect in promoting agricultural production and marketing efficiency. Some of these examples are those of secondary tractor markets in Punjab, contract production of seed and other perishable produce in many states, and new institutions and mechanisms in the credit sector. These kinds of institutional innovations are not only in tune with the new market environment but can also play a significant role in situations where market instruments have not been existing. In the presence of the WTO, new patents regime and MNCs, these institutions can provide a platform for action. The emerging competitive markets in input and output markets necessitate local-level institutions for corrective action. We discuss below some examples of people's institutions in the form of market creation for inputs as well as their management.

INSTITUTIONAL INNOVATIONS IN INDIAN AND PUNJAB AGRICULTURE

There are a large number of institutional innovations in agriculture in India which are being tried by rural producers. These include the Non-Pesticidal Management (NPM) of crops, especially cotton in Punnukula village in Andhra Pradesh that has suffered from excessive use of pesticides on cotton both in terms of economic and physical health of the farmers, who have now decided to make a complete switch over to organic management of cotton and other local crops (CSA, 2004). The regulation of private tube well water prices by village councils in West Bengal is another case of non-market based local-level institutions managing a local private market driven resource (groundwater). Also, in other villages of West Bengal, cooperative tube wells by small and marginal farmers have improved efficiency (lower cost) and equity in water access, and reduced reverse tenancy (Rawal, 2002). Farmers in Kerala have come together in the form of producer companies in undertaking organic production and marketing and in Uttaranchal (Uttarakhand) they have organized themselves into farmers' groups. In Punjab, the second-hand tractor markets are another such case that is discussed in detail below for it is context relevant.

SECOND-HAND TRACTOR MARKETS IN PUNJAB

The local-level people are taking initiative at a significant scale in the case of tractor markets in the Malwa region and more specifically in the cotton belt of Punjab. There are daily, weekly, and fortnightly markets for the sale and purchase of second-hand as well as new tractors in various market towns of the state where thousands of farmers participate in selling or buying tractors. In a state which today has more than 4.5 lakh tractors, there are more than a dozen tractor markets mainly in the cotton belt of the state, which facilitate quick transactions for the buyers and sellers of second-hand tractors. There are different reasons for the sale of old tractors which relate to the larger agricultural economy of the state. The non-viability of tractors due to the small size of landholdings, domestic financial crisis, repayment of bank or other loans, purchase of land, change of model/brand/horse power of the tractor, lack of business for tractors due to their overpopulation and competition, sale and purchase of tractors as a business proposition, and change of occupation are the major reasons for sale of tractors in these markets (Singh and Kolar, 1998). The buyers are satisfied with the operations of these markets as they find a ready market and can liquidate the tractors and other equipment in a short time to meet the exigencies. However, only a small proportion of those who sold tractors bought another one from the tractor market (Singh, 1999).

The operations of these markets are in the hands of a number of informal groups known as *mandis* in each market. They are nothing but collectivities of a few individuals (5–15) who operate as commission agents inside as well as outside the market. These groups generally lease in land for the market, outside the town on an annual basis and share this cost among themselves. This is the major cost. Other costs are working costs for facilities like tents, chairs, and so on. About a few hundred tractors in smaller markets and a few thousand in bigger markets are brought every time (weekly or fortnightly) a market is held. In a small market too, about 20–50 tractors are sold on an average. The model of the tractor and its price are displayed to facilitate a speedy buyer-seller interaction. The commission agents also provide other facilities like space for parking the tractors, and chairs and tents for the farmers. The agents charge commission from both the buyer and the seller, a sum of Rs. 300–500 on each transaction up to Rs. 1 lakh and Rs. 600–1,000 on a transaction above that limit. The payments are made either on the spot or within a week after paying security. The commission agents are responsible for ensuring the payments (ibid., 1999).

Besides the facilitation of business among farmers at one place which lowers search and transaction costs, these markets also generate employment for those who cater to the needs of the people assembled in the market. Further, since these markets supply second-hand tractors, small and medium farmers are able to mobilize money to buy these tractors which are low cost and easily available besides being relevant for these classes of farmers. Recently, even the tractor sales agencies have realized the value of these markets and have started displaying their new models in these markets so that the farmers are made aware of their features and the new product is publicized among potential buyers. These markets are unregulated by any government agency, but they function fairly efficiently so far as the farmers are concerned. Only in one of the markets, the District Collector has allowed the union of these *mandis* to issue licenses (identity cards as commission agents for tractor sales and purchase) to its members. Both the buyers and sellers are made to pay a Red Cross fee of Rs. 50 each per transaction (ibid.).

The above case of innovation of market institution points to the fact that they are institutions of the people, for the people and by the people. And there cannot be a better logic for the existence and success of an institution. It is a case of creation of market infrastructure by groups of local people which facilitates the smooth flow of tractors from sellers to buyers. What seems to lie behind such developments is the local people's initiative and capacity to bring about change whenever required. When there is a felt need, the local entrepreneurs and institutions do come up to provide solutions to the problems unless restricted by some force. In fact, instead of talking about people's participation in government-initiated development projects, one should aim at government or its agencies participating in people's development programmes. It is not surprising to note that many of the above-mentioned people's initiatives are still not known to the government and are not appreciated by its agencies (ibid.).

In the evolution and development of market institutions, we need neither less nor more state but a better state. It is possible that these markets can also lead to some negative developments for certain sectors. For example, it has been observed that sometimes farmers buy tractors with a bank loan and immediately sell them in these markets, as they actually need the cash, not the tractor. This seriously affects the repayment of loan as this money is generally used for purposes other than what the lender intended it for. This kind of market development points to the hidden local potential to create robust local institutions in such a crucial area of economy that is marketing. There is a need to allow and encourage such informal arrangements provided there are no irregularities and practices detrimental to the interests of the market participants. The role of the state in such situations should be supportive and somewhat supervisory in order to ensure the most efficient functioning of these institutions which can deliver objectives of growth, equity, and sustainability (ibid.).

Another important aspect of these institutions is that they function fairly competitively, which is beneficial to the users of these markets as they get fair deals in their transactions. Just as the water suppliers charge very competitive rates as otherwise the user might shift to another supplier, similarly, the tractor market agents charge very nominal commission of 0.3–0.5 per cent of the transaction amount due to the fear that next time or otherwise, the farmer may move to another *mandi*. So, in practice, the commission charged is much lower (Rs. 300 for a transaction up to Rs. 1 lakh) than what is fixed by the *mandi* union (ibid.). The competitive functioning of these markets has meant that they have provided an access to the inputs for small and marginal farmers who otherwise would not have been able to create these facilities on their own (tube wells), or their search costs would have been higher (for tractor purchase). This equity impact of these institutions can be further enhanced if there are supportive public policies, for example, small loans for small farmers to buy second-hand tractors. It is found that the markets came up on their members' initiative, their felt need, and the users have been making use of these markets which function efficiently.

THE FRANCHISEE MODEL

On the output marketing side, contract farming has been seen as an important strategy ever since the diversification attempts began since the late 1980s. And since then contract farming has come a long way in terms of its nature and various experiments with it. The latest 'agri-business facilitators' model of contract farming where processors, banks, and farmers are brought together by an input company is more of a case of 'new business entities' with knowledge and resources and strategy for growth through partnership. They will make money while helping others, including farmers, make money. Their strategies involve bundling of inputs and linking up of credit with input supply which is the agri-business of the twenty-first century (Boehlje et al., 1995). It is interlocking of factor markets coming back in another form. But what is wrong with it if it can provide what the state and cooperatives have not been able to provide for so long, i.e. timely and cost-effective supply of quality inputs and finance and even tractors and combine harvesters on hire basis and assured market for produce? These agri-business services focus on more efficient use of modern inputs with a two-pronged strategy, that is, yield increase or cost reduction through inputs and value addition (market improvement). This is a must for competitiveness, whether domestic or international, where quality and cost effectiveness are the driving forces. Unfortunately, what local panchayats and farmer groups could not do (e.g. custom hiring out of tractors) is being done by agri-business companies. This facilitator model has been now modified with the inclusion of a local *arhtiya* or input dealer as a franchisee for the agri-facilitator (Figure 6.1).

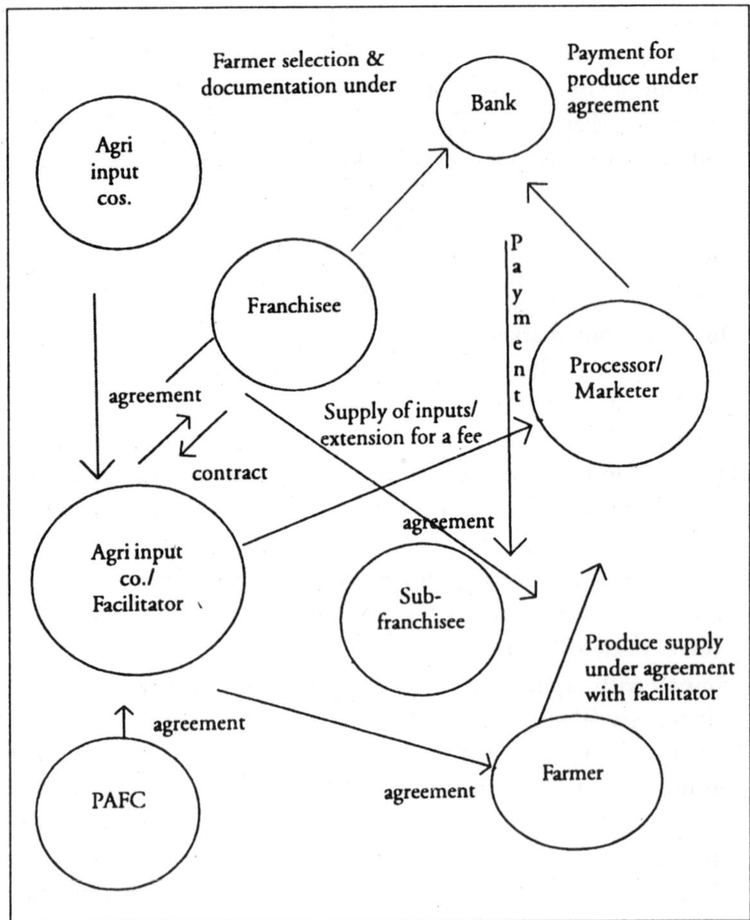

Figure 6.1: The six-partite (networking/franchisee) contrast

But this model does not seem to be working well in Punjab though it has worked well in some other states. In the context of globalization of agriculture, it is important to examine global commodity chains for their small farmer participation as the farmer risk is not being shared even in the networking model of contract farming.

CONCLUSION

A comparative analysis of various marketing arrangements for agricultural produce reveals that it is only the cooperatives and development agencies which can give sustainable benefits for the small and the marginal farmers through market linkages as they are pro-farmer. However for this, they need to receive some policy and financial support from the state (Figure 6.2). They also have to be more market-oriented in their operations which is possible only if they adapt systems of New Generation Cooperatives (NGCs). An NGC is one that has restricted or limited membership, links product delivery rights to producer member equity, raises capital through tradable equity shares among member, enforces contractual delivery of produce by members, distributes returns based on patronage, goes for value addition through processing or marketing, and makes use of information efficiently throughout the vertical system. However, it retains the one member-one vote principle for major policy decisions.

The kind of restructuring helps cooperatives tackle problems of free riding by members, of horizon which is at the root of financial constraint, and that of opportunism, both of the members as well as of the cooperative. This arrangement by cooperatives has helped them become economically efficient, financially viable, and obtain member loyalty wherever it has been tried (Harris et al., 1996). Additionally, non-member procurement and quantity or acreage contracts have also been used by cooperatives to tackle supply side problems (Hinman and Ricks, 1993). Some of the cooperatives like those dealing with sapota (*chikoo*) in south Gujarat have also attempted quality-based grading and pooling system, and contractual relations with members for procurement, along with market

Structure	Sales position of Small Farmers (SF) *vis-à-vis* Large Farmers (LF)	Sales position of SF *vis-à-vis* buyer	Input facilities/ Technical assistance	Government support required
Private local firms	Can be against SFs due to bargaining power	Advantage of access to alternative outlets	May be available based on local experience	Provision of market infrastructure, information, ensuring competition, price stabilization
MNCs/Large firms	Equitable prices through contract	Dependent but secure if supplies quality	Direct supply on credit/direct and intensive	Should negotiate prices and participation for SFs
Cooperatives	Equal if successful	Favourable if efficient coop.	May arrange	Financial support
State boards/bodies	Equal prices if can reach official buying position	May be exploited	Rare/left to other govt. agencies	Insist on reaching small farmers
Development agency	– do –	Protected if meets quality	Direct supply on credit	Financial support required

Source: J.C. Abbott (1993).

Figure 6.2: Relative Benefits of Alternative Marketing Structures for Small Farmers

orientation strategies like multiple outlets and efficient use of market information to achieve better business performance (Singh, 1997).

There is also an emerging lobbying for the removal or relaxation of the Ceiling on Landholdings Act in the state of Punjab which should be resisted as it is neither necessary nor desirable for resolving the present crisis. If operational holdings are to be enlarged for more viable operations, that can be achieved by making the land lease market more efficient or by pooling land together under some co-operative enterprises. If agricultural growth is to be shared in order to realize the virtuous circle of growth and distribution, only a peasant farming system using modern technology of production can achieve it, as the East Asian experience has shown. Not only is it more competitive compared to the capitalist/corporate farming system, but also peasants respond and adopt new technologies of production whenever the opportunity arises. The experience of the Green Revolution in Punjab is an excellent example of this. Secondly, it is able to employ more labour than capitalist farming can ever do, given its normal motive is to maximize profit (Mishra, 1997). The peasant farming system under the new technology regime also helps employment generation for rural labour as the peasant farmers substitute labour for capital in a much better way.

Given the nature of modern farming involving tremendous amount of technological input and market orientation which require capital resources, it is but inevitable to involve private corporate business interests in agricultural development through the contract farming system. What is required is marketing extension in terms of better product planning at the farmer level, provision of market information, securing/accessing markets for farmers, provision of alternative markets and market orientation in terms of improved marketing practices at the farmer level (Patnaik, 2003). The Punjab government has also been now resigned to a role of a facilitator of contract farming in the state. The governments of Uttar Pradesh and Punjab have recently amended the Agricultural Produce Market Committee (APMC) Act which did not permit farmer-level (direct) procurement by companies. This legal reform process is being accelerated by the central government with the enactment of the Model Act for the state Agricultural Produce Marketing (Development and Regulation) Act, 2003 which deals with the setting up of private markets, selling of produce by growers outside the APMCs (regulated markets), setting up of direct markets, specialized commodity-specific markets, regulation and promotion of contract farming, provision for agencies and measures to promote quality, standards, and alternative markets, and public-private partnerships to facilitate more and better linkage between firms and farmers (GoI, 2004). The amended APMC Act has certain mandatory and optional provisions wherein mandatory ones include aspects like who can undertake contract farming, contract specifications, liabilities, farmer asset indemnity, and dispute resolution. On the other hand, the optional features include those relating to farm practices, insurance, monitoring, role of farmer body, and support from the sponsor. Under this Act, registration for contract operations with the local authority is a must and the model contract agreement is quite fair in terms of sharing of costs and risks between the sponsor and the grower (GoI, 2003).

There is a role for state agencies and NGOs to intervene in contract situations as intermediaries to protect the farmer and the broader local community interests. The NGOs can also play a role in information provision, and in monitoring and regulating the working of contracts. Better cooperation and coordination between companies and cooperatives for agricultural development also needs to be encouraged. Further, both companies and the state should promote group contracts with the intermediation of local NGOs and other organizations and institutions so that contractual relationships are more durable, enforceable, and fair. An insurance component in farming interventions is a must to protect farmer interests and it is noted that some companies are already doing it. But the most important thing is to ensure market for the farmer produce at better prices under these agri-business projects. The government should also play an enabling role through legal provisions and institutional mechanisms, like helping farmer cooperatives and groups, to facilitate smooth functioning of contract system.

It is important to realize that for diversification to be meaningful it must go beyond crops into allied and non-farm enterprises.

Further, the state's farm sector problems like high cost of farming, low returns, and environmental sustainability require different kinds of institutions (collectivities instead of individual enterprise) for which institutional innovations are a must. Punjab has also not tried the new organizational option made available by policy change, i.e. the producer companies (cooperative companies) under the Companies Act which farmers in many states have gone ahead with in various existing and new projects.

REFERENCES

Abbott, J.C. (1993), 'Marketing, The Rural Poor and Sustainability', in J.C. Abbott (ed.), *Agricultural and Food Marketing in Developing Countries: Selected Readings*, Oxon (UK): CAB International, pp. 65–92.

Alagh, Y.K. (1990), 'Agro-Climatic Planning and Regional Development', *Indian Journal of Agricultural Economics*, 45 (3), July–September.

——— (1982), 'Indian Agricultural Economics—Some Tasks Ahead', (Inaugural Address at the 42nd annual conference of the ISAE held at Pantnagar, *Indian Journal of Agricultural Economics*, 37 (4), October–December.

Batra, S. and A. Singh (2003), *Evolving Proactive Power Supply Regime for Agricultural Power Supply*, Management Traineeship Segment (MTS), Report no. 1562, Anand: Institute of Rural Management.

Baumol, W.J. (1990), 'Entrepreneurship: Productive, Unproductive, and Destructive', *Journal of Political Economy*, 98 (5).

Bhatia, B.M. (1987), *Towards a Solution of Punjab Problem: Restructuring and Diversification of State Economy*, New Delhi: Centre for Policy Research.

Biswas, U.N. and S.N. Biswas (1997), 'Pro-economic Attitudes and Development: A Cross-cultural Study', *Psychology and Developing Societies*, 9 (2).

Boehlje, M.J. Akridge and D. Downey (1995), 'Restructuring Agribusiness for the 21st Century', *Agribusiness*, 11(6).

Chand, R. (1999), 'Emerging Crisis in Punjab Agriculture: Severity and Options for Future', *Economic and Political Weekly*, 34(13), 27 March, A2-A10.

CSA (Centre for Sustainable Agriculture) (2004), *No Pesticides, No Pests: The Remarkable Story of How a Village in AP Rid Itself of Pesticides*, Hyderabad: CSA, October.

Gandhi, V. P. (1997), 'Technology, Cost Reduction, and Returns in Agriculture: A Study of Wheat and Rice in Punjab', *Vikalpa*, 22 (2), April–June, pp. 35–48.

Gill, S.S. and J.S. Brar (1996), 'Global Market and Competitiveness of Indian Agriculture: Some Issues', *Economic and Political Weekly*, 31(32), 10 August, pp. 2167–77.

Government of India (2003), *Contract Farming Agreement and its Model Specifications*, Ministry of Agriculture, Department of Agriculture and Co-operation, New Delhi: Krishi Bhawan, 17 September.

——— (2004), *Economic Survey 2003–4*, New Delhi: Ministry of Finance.

Government of Punjab (2006), *Statistical Abstract of Punjab*, Chandigarh: Government of Punjab.

Harris, A., B. Stefanson and M. Fulton (1996), 'New Generation Co-operatives and Co-operative Theory', *Journal of Cooperatives*, vol. II.

Hinman, L.D. and D.J. Ricks (1993), 'Co-operatives, Marketing Orders and the Coordination of Supply and Demand for Perennial Crops in the U.S.', in Csaba Csaki and Yoan Kislev (eds.), *Agricultural Co-operatives in Transition*, Boulder: Westview Press.

Mishra, S.N. (1997), 'Agricultural Liberalisation and Development Strategy in Ninth Plan', *Economic and Political Weekly*, 32 (13), 29 March.

Patnaik, G. (2003), 'Marketing, Storage, and Extension Services: State of Agriculture in India', in B. Debroy and A.U. Khan (eds.), *Enabling Agricultural Markets for the Small Indian Farmer*, New Delhi: Bookwell, pp. 81–120.

Rangi, P. S. and M.S. Sidhu (2000), 'Problems and Prospects of Agriculture in Punjab', in R.S. Bawa and P.S. Raikhy (eds.), *Punjab Economy: Emerging Issues*, Amritsar: Guru Nanak Dev University.

Rawal, V. (2002), 'Non-Market Interventions in Water-Sharing: Case Studies from West Bengal, India', *Journal of Agrarian Change*, 2(4), October, pp. 545–69.

Sidhu, B.S. et al. (1998), 'Sustainability Implications of Burning Rice and Wheat Straw in Punjab', *Economic and Political Weekly*, 33 (39) 26 September, A163–A168.

Sidhu, H.S. (2002), 'Crisis in Agrarian Economy in Punjab: Some Urgent Steps', *Economic and Political Weekly*, 37 (30), 27 July, pp. 3132–8.

Shah, C.H. and C.N. Vakil (1979), *Agricultural Development of India: Policy and Problems*, Bombay: Orient Longman.

Shah, T. (1995), 'Institutional Dimensions of Agricultural and Rural Development in India', *Administrator*, 40 (1).

Singh, J. and J.S. Kolar (1998) 'Why Farmers are Forced to Sell Tractors?', *The Tribune*, Chandigarh, 6 June.

Singh, K. and S. Kalra (2002), 'Rice Production in Punjab: Systems, Varietal Diversity, Growth and Sustainability', *Economic and Political Weekly*, 37 (30), 27 July, pp. 3139–48.

Singh, L. and S. Singh (2002), 'Deceleration of Economic Growth in Punjab: Evidence, Explanation and a Way Out', *Economic and Political Weekly*, 37 (6), 9 February, pp. 579–86.

Singh, S. (2000), 'Crisis in Punjab Agriculture', *Economic and Political Weekly*, 35 (23), 3 June, pp. 1889–92.

——— (1999), 'Institutional Innovations in Indian Agriculture: A Case of Input Markets', *Institutional Development*, 6(2), November, pp. 3–10.

——— (1999a), 'Green Revolution and Agro Industrialization: A Case Study of Primary Food Processing Industries in the Indian Punjab', *International Journal of Punjab Studies*, 6 (1), pp. 55–83.

——— (1997), 'Managing Marketing: Co-operatives in Horticultural Business in South Gujarat', *Indian Journal of Agricultural Marketing*, 11 (1&2).

CHAPTER 7

Corporate Agriculture Farming in West Punjab: A Serious Threat to Small Peasants

Zia ur Rehman

INTRODUCTION

Like most developing countries, majority of Pakistan's population is in rural areas and involved more or less in subsistence agriculture. Traditionally, agriculture has not been just an economic activity but a way of life; the farmers' socio-cultural values and emotions are rooted in agriculture. But disregarding traditional agriculture, modern agricultural technologies were introduced in the 1960s and further boosted in the 1980s. The way these technologies are being introduced and agriculture commercialized, rural communities are losing subsistence agriculture as well as their socio-cultural values. It is now widely recognized that technology has benefited a few of the big landlords and, particularly, corporations involved in agri-business. As a result, the many farmers are gradually losing their hold on agriculture since they cannot afford the increasing costs of modern inputs to compete with the big landlords. Consequently, an alarming number of villagers are leaving their agricultural activities and migrating to and settling down in urban centres. In contrast, corporatization has grown fast, particularly under the World Trade Organization (WTO) regime. The MNCs have become so powerful that they are actually controlling the world's resources and power. The budget of one MNC is sometimes greater than the collective budget of many developing countries. Today, the MNCs are playing havoc in the business of seed and agro-chemicals. Under these circumstances it is becoming increasingly difficult for governments to protect the rights of their farmers.

In the light of these facts, the recently announced corporate farming plan of the Government of Pakistan seems unrealistic. It may, in fact, prove to be the last nail in the coffin of subsistence farmers who are already economically hard pressed. This would be a big alteration in the structure of agriculture historically practised in the country since, with the exception of a few large landowners, about 94 per cent farmers in Pakistan are small landowners and tenants.

SUSTAINABLE AGRICULTURE *VERSUS* CORPORATE FARMING

Due to rising environmental awareness sustainable agriculture is gaining importance. Keeney (1989) defined sustainable agriculture as 'agricultural systems that are environmentally sound, healthier, productive and that maintain the social fabric of the rural community'. Recently the emphasis has been on harmonizing agriculture with the natural ecosystem. Thus, sustained agricultural development requires careful use and monitoring of quality of the natural resource base in order to avoid its deterioration.

The society, however, is increasingly putting more pressure the on agriculture sector to meet the food and fibre requirements of a rapidly growing population. To fulfil this growing demand, the agriculture sector must grow at a rate reasonably higher than that of the population growth. The policy of the Government of Pakistan so far has been to increase productivity through increasing efficiency by using high yielding varieties, more chemical inputs and heavy farm machinery. However, following this practice the production of grain cannot increase indefinitely. Therefore, the Government of Pakistan is now going for the genetically modified seeds. There are two main reasons why this policy is unrealistic and unsustainable. On the one hand, it is being argued that the introduction of modern agricultural inputs has not solved the problem. Table 7.1 shows that with the increased use of modern

TABLE 7.1: AVERAGE YIELDS OF IMPORTANT CROPS IN PAKISTAN

	Wheat (kg/ha)	Rice (kg/ha)	Maize (kg/ha)	Sugar cane (tonnes/ha)	Cotton (kg/ha)
1965–70	987	1,162	1,079	33.2	286
1970–5	1,217	1,531	1,153	35.7	341
1975–80	1,848	1,571	1,249	37.0	286
1980–5	1,596	1,685	1,267	37.5	343
1985–90	1,773	1,600	1,340	39.7	544
1990–5	1,951	1,626	1,410	44.1	596
1995–2000	2,196	1,922	1,658	47.0	557

Source: GOP: *Agricultural Statistics of Pakistan* 1978, 1984, 1989–90 and 1999–2000.

agricultural inputs, the crop yields improved in the beginning but have stagnated for the last few years although the application of fertilizers and pesticides has increased manifold. For instance, the consequences of overuse of pesticides are evident in the cotton growing areas of Pakistan, which uses almost 70 per cent of the total pesticide use in the country. The statistics confirm that increased use of pesticides has neither solved the pest problem nor led to higher cotton production. Figure 7.1 shows that with the increased use of pesticides, cotton production increased in the beginning, became stagnant after a while and has started declining over the last few years.

On the other hand, it is also a well recognized fact that over time, the increased use of modern agricultural inputs have degraded the environment besides generating numerous other serious side-effects on man and nature. The use of fertilizers and pesticides is polluting the soil, ground water and air. Similarly, due to use of heavy machinery for intensive cultivation, the topsoil is disappearing at an astounding rate of 25 billion tonnes a year all over the world. In the 1700s, America had a base of 21 inches of topsoil. Today there are only about six inches of topsoil in most of its areas. Corporate farming is largely held responsible for this problem. For every inch of topsoil lost, the grain yields drop by 6 per cent.

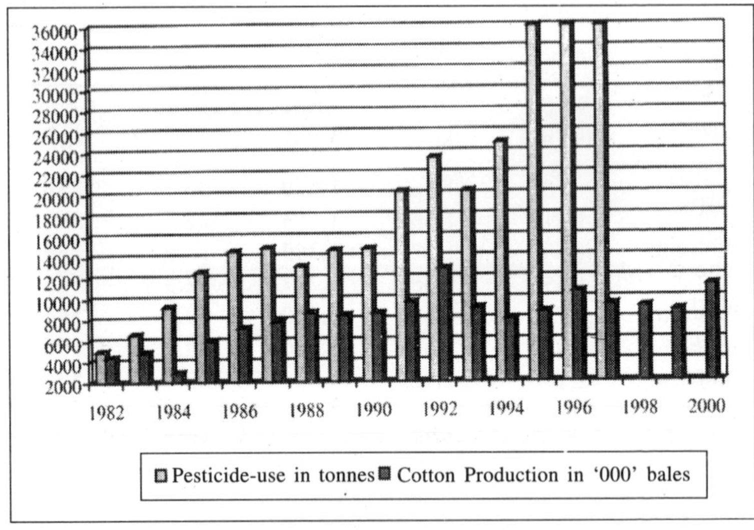

Source : Hasnain, 1999.

Figure 7.1: Cotton Production *versus* Pesticide Use

Thus, with the growing awareness of the consequences associated with modern agricultural inputs, it is becoming increasingly difficult to increase or even continue with the present levels of their use indefinitely.

After going through these examples, sustainable agriculture appears to be the only lifelong and eco-friendly way of agriculture which has the potential to take care of the subsistent farmers, reduce their poverty, and ensure food security and sustainable development of the country. As a first step towards this, there is a need to seek ways to reduce dependency on modern agricultural inputs through better crop management strategies on the one hand and enhance crop productivity to provide adequate food and fibre for a rapidly growing population on the other. This poses a new challenge to our national research systems. They are now required to generate technologies that are not only compatible with our socio-economic systems and cultural values but also capable of enhancing crop yields. And this it should be done in environment-friendly ways. The overall motive should be that the present and future generations should have equal access and rights to use the natural resources for economic and social well-being.

Unlike sustainable agriculture, where the strategy is to minimize environmental and economic risks by using management and labour skills to diversify production and conserve natural resources, corporate farming encourages large-scale, highly mechanized and specialized farms where uniformity is emphasized over quality and where decisions are solely profit oriented. Instead of taking care of the natural resources to be safely handed over to the coming generations, agri-business corporations in their attempt to earn more in a short time, use modern technologies like heavy machinery, high yielding and genetically modified varieties of crops, high doses of chemical fertilizers and pesticides. The produce—grain, fruits and vegetables—are also subjected to post-harvest chemical treatment for better marketing. The big businesses don't just stop there. They also monopolize marketing systems and influence the country's laws to promote their businesses. Listed below are some social and ecological costs of corporate farming as experienced by many developing countries, particularly, with regard to its effects on the rural poor/small farmers and food security.

1. Agricultural corporatization and production for the export market leads to the uprooting and destruction of small farmers while benefiting the large farmers. Agricultural corporatization and exports increase single commodity harvests. When all farmers are growing the same commodity over large areas, the prices come down, while the costs of inputs that are imported are ever escalating. As a result, the farmers' profits fall drastically. As cost of production increases the farmers experience a cost-price squeeze. In this process, only the large farms can survive (Rosset et al., 2000).
2. The logic of the market is such that the costs for small farmers to use high-tech input systems are larger than for the big farmers. The poor farmers cannot afford to buy the fertilizers, pesticides and other inputs in bulk. The big farmers get discounts for large purchases. The poor farmers also cannot hold out for the best price for their crops, while the larger farmers whose material conditions are sound wait for the right time. Big farmers can

afford to pay for irrigation services, which may not be within the reach of small farmers.

3. The increase in yield from high-yielding Green Revolution technologies have been decelerating, and in some cases stagnating and even contracting. The highest yields have been obtained by using ever larger inputs of fertilizers and irrigation water, which in many places have passed the point of diminishing returns. Greater use of these inputs has become less productive (Altieri and von der Weid, 2000). In Philippines, rice production increased in the late 1970s and early 1980s as a result of the Green Revolution, but has since been on the decline. Analysts attribute this to the very nature of these high-yielding varieties (Mariano, 1996). In comparison to the traditional varieties, outputs in terms of energy are also small. For instance, traditional rice farming in Asia produced 10 times more energy in food than was expended to grow it (Lappe et al., 1998). Today's Green Revolution rice production cuts the net output to half. According to the Cambridge University geographer, Bayliss-Smith, the gains drop to zero in a fully industrialized system such as that of the US (Bayliss-Smith, 1984). Why is this happening? Experts conclude that the chemical fertilizers have destroyed the quality of the soil. Yields are falling because the chemical fertilizers are not a sustainable source of soil fertility (Rosset, 1996). In the long run, these high-tech methods encourage desertification, soil erosion, pesticide contamination and the depletion of groundwater (Lappe et al., 1998). Yet these ecological problems are ignored because of the difficulty in quantifying and assigning monetary values to ecological degradation. These remain hidden costs behind the yields obtained in conventional farming.
4. One of the myths about industrial agriculture is that it is a land-saving system. In fact, monocultures and the separation of crops from livestock necessitate much higher amounts of land than traditional poly-cultures and systems that integrate crops and livestock.
5. The monocultures eliminate diversity and the multiple uses of land otherwise necessary for an integrated farm unit. Under

such conditions of monoculture cropping additional acres are required to produce those products that could otherwise have been produced on the same piece of land.

6. Shiva explains it as follows: If traditional varieties produce 1,000 kg/acre grain and 1,000 kg/acre fodder, then industrial varieties produce 1,200 kg/acre of grain and zero kg of fodder. Thus another acre has to be used for fodder production. The 2 acres, without chemical inputs and new seeds, could have produced 2,000 kg grain plus 2,000 kg of fodder on 2 acres (Shiva, 1999).
7. Likewise, crop production strategies which ignore livestock as an important source of soil fertility lead to an intensification of agriculture through the use of non-sustainable chemical inputs like fertilizers, weedicides and pesticides.
8. In addition to it, industrial livestock farming in fact consumes three times more acres of land than ecologically rearing livestock. Analysts have calculated that Europe, in fact, uses seven times the area used in the Third World countries for cattle feed production. For fodder alone (including that used to produce food products for export) the Netherlands appropriates 100,000 to 140,000 sq km of arable land. This is 5–7 times the area of land under agriculture in the entire country (Wackernagel and Rees, 1998).
9. It has been concluded that the combination of industrial plant breeding and industrial animal breeding increases the pressure on land use by a factor of 400 per cent while separately increasing output of grain and milk by only a factor of 20 per cent (Shiva, 1991).
10. Another resource which is overused by industrial agricultural methods is water. As mentioned earlier, HYV seeds grow well only when water is plentiful. The water tables, however, are falling in key food producing countries. About 480 million of the world's 6 billion people are fed with grain produced by over-pumping aquifers (Brown, 2000). This is already creating serious water shortage in various countries today. Regions that are suffering from aquifer depletion include the central and northern China, north-west and southern India, some parts

of Pakistan, much of the western United States, North Africa, the Middle East and the Arabian Peninsula (Postel, 2000). Besides constraining future food supply, the groundwater over-pumping is widening the income gap between the rich and poor. The poor cannot afford to deepen their wells or buy powerful pumps. As the shallower wells dry up, some of the small-scale, poor farmers end up renting their land to the deeper well owners and become labourers on their farms. The agronomic systems of agricultural production, like inter-cropping to maximize use of soil moisture as well as better matching crops to climate conditions can reduce the pressure on water supply and increase the opportunities for small farmers to maintain their livelihoods (ibid.).

11. The point made here is that extra resources used by the industrial systems, such as the Green Revolution techniques, could have been used to feed people. But these systems are resource wasteful and rely on intensive external inputs, including water and land. The emerging genetic revolution too (which is similarly input intensive) is resource wasteful. These systems take away the entitlements from those most in need, and add to food insecurity and poverty in the developing countries (Shiva, 1999).
12. Another significant issue that must be addressed as a result of industrialized food production systems is the increasing dependence on pesticides. Monocultures erode diversity, but it is diversity that increases resistance to pests. In a holistic system, a problem in a part of the system can be absorbed or counterbalanced by the entire system. This is why the poly-cultural system of cropping is widely recognized as more stable. This is not the case with monocultures. Without the advantages of a balanced eco-system, pest infestations are so common that they may destroy the entire crop easily.
13. Chemical pesticides are in fact creating more pests than controlling them. By contributing to the emergence of resistance in pests and by destroying their natural enemies and predators, pesticides in fact increase the pest problem. Globally, 450 species of insects and mites, 100 species of plant

pathogens and 48 species of weeds have become resistant to one or more pesticide products. In California, of the 25 insect pests, 72 per cent are resistant to one or more pesticides. Of these, 96 per cent are either created or aggravated by pesticides themselves (ibid.).

14. The strengthening and growing resistance of pests is hardly surprising given that 4.7 billion pounds of pesticides are used annually throughout the world (Natural Resource Defense Council [NRDC], 1996). The world pesticide use has increased twenty-sixfold in the last 50 years. In 1996, total pesticide sales topped $33 billion. The value of world exports of pesticides increased more than eight times in less than four decades, from $1.3 billion in 1961 to $11 billion in 1997. Most of these exports originate from the industrial countries (McGinn, 2000).
15. The other facet of the pesticide problem is the cost to human health. Estimates of pesticide poisonings in the Third World are as high as 25 million people yearly despite the fact that many cases go unreported (Jeyarathum, 1990).
16. The most harmful chemicals end up for sales in the developing countries. Many industrialized countries continue to export chemicals that are deemed too deadly for domestic use. Despite the awareness of bio-accumulating chemicals (that is, each link or species in the food chain takes up the previous link's exposure, adding it to their own and magnifying the effects), this trend is getting worse. Between 1992 and 1996, the US chemical companies increased exports of domestically prohibited pesticides by 18 per cent. The rate of exports of pesticides that have never been registered jumped the 40 per cent mark during this time (McGinn, 2000). More than 14 per cent of all the US pesticide exports are in some way restricted from use within the country itself (Carl, 1998).
17. Besides causing cancer, new evidence suggests that pesticides may have many other dangerous effects on human beings. For example, many pesticides fall into a category of chemicals called 'endocrine disrupters', some of which directly affect the reproductive system (Bouque and Bekkers, 1997).[1]

18. Much of the problem with the use of pesticides has to do with meeting the cosmetic standards of consumers in the export markets, particularly the industrialized countries. Lappe, Collins and Rosset estimate that in the US, 60–80 per cent of the pesticides applied to oranges and 40–60 per cent applied to tomatoes are meant only to improve their appearance. Less than 0.1 per cent of the pesticides applied to crops actually reach the targeted pests. The rest moves into the ecosystem, contaminating the land, water and air (Lappe et al., 1998).

CORPORATE AGRICULTURE FARMING PLAN

The Government of Pakistan has recently announced a Corporate Agriculture Farming (CAF) plan following the policy to boost the country's economy. The federal Ministry of Food, Agriculture and Livestock (MINFAL) has identified specific agriculture sectors with a view to attract foreign as well as local investors. The objective of this step is to commercialize the agriculture sector. Besides identifying the harnessing of cultivable wasteland areas, the government also pinpointed fisheries sector, production of mutton through raising of sheep and goats, dairy farming, dehydration of vegetables, off-season vegetable production, animal feed mills, fruit juice making plants, solvent oil extraction from rice bran, tomato paste production and sunflower hybrid seed production.

Proposals relating to foreign investment in the agriculture sector have been processed by the Board of Investment (BOI) in consultation with respective provincial governments and approved by the competent decision-making authority. In this connection, an investment policy for the agriculture sector was also announced by the BOI recently under which the import tariff on agricultural machinery (not manufactured locally) will be zero. According to the BOI, the detail of the CAF plan is as follows:

Objectives

- Large local market increasing at 29 per cent per year.
- Internationally competitive unit cost of production for all major crops, fruits and vegetables.

- High quality agricultural products due to favourable resource base.
- Low transportation cost and developed routes to the Middle East, Iran, Afghanistan and the Central Asian countries.
- International quality inputs available in the country.
- Market driven policies.
- Least intervention from the government.

Most Liberal Investment Policy

- CAF is declared an industry but with the exemption that it would not be subject to those levies and industrial development laws that are applicable to an industry, after 3–5 years.
  - 100 per cent foreign equity allowed.
  - No minimum foreign investment.
  - Allowing remittance of capital, profits and dividends.
  - CAF will enjoy credit and other facilities.
  - Allowing the local or foreign, private or public limited companies to invest in corporate farming that are locally incorporated under the Companies Ordinance, 1984.
  - No upper ceiling on landholdings.
- State land can be either purchased or leased by the foreign or local companies for a period of 50 years that is extendable for another 49 years. The foreign company, however, cannot transfer such land to other foreign company without the permission of the federal and the concerned provincial governments.
- All banks and financial institutions will earmark separate credit share.

Attractive Fiscal Incentives

- Zero per cent customs duty on the import of agricultural machinery, equipment and implements—new or used—that are not manufactured locally.
- Government of Punjab has exempted from stamp duty and registration fee all transfers of land in favour of agricultural companies registered under Companies Ordinance, 1984. The

Government of NWFP has also provided one-time exemption of stamp duty.

- Tax relief: First Year Allowance of 75 per cent of PME cost.
- In addition:
  - Provincial Agriculture Income Tax (AIT) laws provide for proportionate liability of corporate shareholders.
  - Dividends from corporate agricultural farms (for non-industrial activities) shall not be subject to tax.
  - Farm income is given more favourable treatment than non-farm corporate incomes because of the risk/uncertainty associated with farming.
  - Existing definitions of farming activity, as distinct from processing/industrial activity will continue to be maintained.

Investment Opportunities

- Land development/reclamation of barren, desert and hilly land for agriculture purposes.
- Reclamation of water front areas or creeks.
- Crops, fruits, vegetables, flower farming/integrated agriculture (cultivation and processing of crops).
- Modernization and development of irrigation facilities and water management.
- Forestry.
- Horticulture.
- Dairy farming.
- Livestock farming, breeding and small ruminants (sheep, goat).
- Production of quality seeds.
- Fruits, vegetables and flowers—grading, processing, packaging and preservation, etc.
- Seafood (farming/fishing, processing and preservation of fish, shrimps and other marine products).
- Storage facilities for agricultural produce.
- Marketing and export of agricultural produce.
- Cool chains (separate package).

DRAWBACKS OF THE CAF PLAN

The promoters of corporate agriculture are not basically different in operation from other multinational corporate enterprises. The corporate greed for profit from renewable natural resources would lead to growing food insecurity and prove to be the last nail in the coffin of subsistence farmers who are already economically hard pressed. The proponents of the corporate farming system have themselves defined that in this system the farm owner, the manager and the farm worker are different people. Introduction of this system of agricultural production would be a major blow to the age-old agricultural practices prevalent in South Asian countries in which the farming labour is also empowered to make decisions and reap profit. Experts have already expressed concern that the small farmers who are 94 per cent in Pakistan's agriculture sector, having 37 per cent of the land, would be the major sufferers after the implementation of this scheme. However, the big landlords holding more than 60 hectares of land, constituting only 6 per cent of the farming community but controlling 63 per cent of the total land will not be affected in any way by corporate farming.

Even prior to the corporatization of agriculture in Pakistan, the market forces are already determining the prices of different agricultural commodities. These forces have made the small farmers tools in their hands. Recently, the ruthlessness of the market forces have been witnessed by the wheat and rice growers who were not able to sell their produce at the support price and were forced to burn heaps of their yields in the local grain markets. Under such conditions, how can the small peasants compete with the financial, marketing and political clout of giant corporations?

These corporations will swallow the land of the small peasantry and make them redundant for farming since they will not be able to make both ends meet. They would be forced to find alternative sources of livelihood. It would increase mass displacement of rural population and its movement towards urban centres, resulting in increased haphazard urbanization. That would not only worsen the pressure on urban resources, but also hundreds of acres of prime agricultural land around the urban centres would be taken up to accommodate the displaced people. The prime objective of the proposed CAF plan in Pakistan is to combine agricultural and processing activities at one place to produce high-quality agricultural products. This involves the use of latest production technology, adequate use of inputs like fertilizers, pesticides, better quality seeds, heavy machinery and efficient use of water.

The objectives, no doubt, appear to be very impressive but can technology or high technology really work by itself? The option to increase productivity by using HYVs, more chemical inputs and heavy farm machinery has also negative side effects on both man and nature. It has already been discussed in the last section that with excessive use of modern agricultural inputs, the crop yields increased in the beginning but subsequently became stagnant. It may, therefore, be inferred that technology by itself cannot fix it all. We need to think of sustainability as well.

The problem of labour would be more serious under the corporate farming plan since labour laws would not be applicable to corporate agriculture companies, thus depriving labour of its voice and rights. Moreover, corporate farming will rely heavily on modern equipment since there is an incentive to import new or used agricultural machinery, equipment and implements, etc., at zero per cent customs duty. The mechanization may result in reduced production costs but it would also reduce employment for labour. This would further aggravate the already existing problem of mass unemployment in Pakistan.

On the basis of above-mentioned reasons, the civil society opposes such an understanding of the government that everything can be bought and sold in the market and that efficiency is the highest good, and profit the highest goal. This kind of thinking deeply offends humane society. We have to save our agriculture and the farming community from MNCs by protecting our indigenous systems of knowledge. We have to promote organic and cooperative farming instead of corporate farming. This alone will lead to sustainable development. Corporate farming does not suit us given the human and natural resources at our disposal.

UPDATE ON CORPORATE AGRICULTURE FARMING

The year 2004

- Cabinet approves the 'policy package of CAF' submitted by the MINFAL along with a time-bound plan subject to the modification that CAF would not be declared an industry.
- Cabinet decides that the Ministry for Labour, Manpower and Overseas Pakistanis, Ministry of Industry and Production, and MINFAL would prepare a labour package for CAF in five years.
- Cabinet proposes an amendment in the Land Reforms Act, 1977 keeping in view the CAF package:

'Person' includes a religious, educational or charitable institution, every trust whether public or private, a Hindu undivided family, a company or association or body of individuals, and a cooperative or other society, but does not include *an agricultural farming company, incorporated under the Companies Ordinance, 1984 (XLVII of 1984), and duly listed on the stock exchange,* a local authority, a university established by law, a joint stock company which is directly or indirectly held or controlled by Federal Government or Provincial Government or by both or any other body in which the Government holds majority interest.

DEMANDS BY CIVIL SOCIETY ORGANIZATIONS

In the year 2004, after the cabinet approval of CAF, the civil society groups and the farmers' organizations have conducted massive rallies in different cities expressing their concern in this regard to the Government of Pakistan. The unemployed graduates in agriculture in Quetta city sat on hunger strike against this policy of the government and demanded that the land which is being given on lease to foreign companies should be given to them and other farmers. This would not only provide them gainful employment, but also ensure sustainable development of both agriculture and economy.

In their press conferences, press releases, and demonstrations, the farmers and civil society have demanded that the government should also consult small farmers and civil society organizations while taking any decision on corporate farming. They have also

CAF ACTION PLAN

	Policy package elements	Implementation responsibility
1	*Corporate Agriculture Farming (CAF) as an Industry:* Government may declare CAF as an industry along with the exemption that it would not be subject to those levies and industrial establishment laws, which are applicable to an industry, after 3-5 years.	Ministry of Industries and Production/ BOI/MINFAL
2.	*Legal entity:* Only such local and foreign companies will be entitled to CAF that are locally incorporated under the Companies Ordinance, 1984.	Security and Exchange Commission of Pakistan & Provincial governments
3.	*Land ceiling:* There may not be any upper ceiling on land holding for CAF by amending relevant laws. The size of the proposed corporate farms may be left to be determined by the prospective investor.	Federal Land Commission
4.	*Taxation:* AIT Regime applicable in provinces on incomes from agriculture would be applicable to CAF. Tax relief in the shape of first year allowance of 75 per cent of machinery cost would be allowed to set-off provincial AIT.	Provincial revenue departments
5.	*Labour laws:* Labour laws may not be presently applicable to corporate agriculture companies. Due to special circumstances of the agriculture sector, however, appropriate labour laws be developed for this sector within five years.	Ministry of Labour and Law, MINFAL

6.	*Import duties:* 1. Zero-rated duty to be charged on import of agriculture machinery and equipment and also exempted from sales tax. 2. Machinery items for wheat/grain storage and cool chain may be included in SRO 437/(1)/2001.	CBR
7.	*State land:* Wherever possible, state land may either be sold or leased to the investor for 50 years extendible for another 49 years. Preference in this respect will be given to culturable wasteland, which is otherwise fit for cultivation.	Board of revenue, provincial governments/BOI
8.	*Duty on transfer of land:* Government of Punjab has exempted from stamp duty and registration fee all transfers of land in favour of agricultural companies registered under Companies Ordinance, 1984. Government of NWFP has also provided one-time exemption of stamp duty for initial purchase/lease of land to CAF. Other provinces may follow suit.	Provincial revenue departments

criticized the government for making a U-turn on land reform polices and selling state land to corporate companies, ignoring the rights of small farmers.

CONCLUSION

It is needless to say that the majority population in rural Pakistan depends on agriculture. It is basic to our economy. It is a vital sector that cannot be ignored. If this is correct, it is recommended that agriculture should not be subjected to experimentation without proper planning on the part of the government. The side effects of the new agricultural policy should duly recognized. Imagine for a while, what will happen if agriculture goes out of our hands? We will be dependent on other nations in the short run and may not be able to defend our own country. Over the past fifty years, we have already lost a lot of our agricultural wealth. We imported Green Revolution technology (a technology invented in the West for their own needs) without considering its impacts on agriculture and rural communities. As a result, we have lost our traditional agriculture, one that had been refined over the centuries, and in the process we have disrupted our socio-economic and cultural systems. Over time, few other menaces such as fertilizers and pesticides have become compulsory inputs for agriculture that further deteriorated it. As a matter of fact, this type of farming was encouraged and protected by the feudal landlords. Unluckily, throughout the history of Pakistan, the government has always been dominated by the feudal lords. They, in one way or the other, have always managed to remain in power and always succeeded in moulding the law of the country in their favour. They have never allowed land reforms, which are vital for just development. The current corporate farming plan is another selfish move on their part to eliminate the remaining chances of land reforms.

At the moment the feudal lords are not alone. Earlier, the East India Company flourished with them in the subcontinent from 1857 to 1947. Likewise, now agri-business MNCs are their equal partners. Recently 19 national companies have registered themselves for CAF (see details on the next page). The MNCs are playing havoc with the developing countries by influencing country laws and looting its resources.

RECOMMENDATIONS

- Strict and justifiable land reforms should be implemented without further delay so that the farming community owns agriculture and comes up with desired agricultural production, leading to prosperity at the local as well as the national level. Small farmers, in fact, have a proven record in raising agricultural productivity as well as employment.
- There is need to deal carefully with agriculture due to the fact that this is the vital sector on which the whole of Pakistan depends. Adoption of a wrong agricultural policy will damage our country's future.
- To ensure sustainable development, in the exiting structure of public governance needs to be reconsidered. Evidence shows that lack of political commitment, inability to tackle with international pressure, weak institutional structure, inadequate delivery systems, indifferent staff attitudes, vested interests, financial leakages and malpractices have undermined the effectiveness of government programmes. The impact of government programmes has not been proportionate to the human and financial resources employed. There is need to harmonize the Government of Pakistan's policies across all ministries and sectors. At the moment, some strong ministries propagate policies which sometimes contradict with those of others. For instance, the idea of corporate farming is an attractive one to the Ministry of Commerce and Finance but for the Ministry of Agriculture, it cannot be a good idea because it will ruin the traditional agricultural system and kill the hope of practising sustainable agriculture in the country. Similarly, the environmental problems and displacement of rural communities and pushing them towards cities as destitutes are the concerns of the ministries of environment, labour, local government and rural development. The Ministry of Health will also be not happy due to intensive use of fertilizers and pesticides. Yet, the idea of corporate farm-

Company Name	Incorporation	Registered office & address	Sectoral classification
1. Farms (Pvt.) Ltd.	25 Apr 83	Lawrence Road, Lahore	Corporate Agriculture Farming
2. Greenacres Farms (Pvt.) Ltd.	13 Nov 80	H #31, St. 52, F-7/4, Islamabad	—
3. Iffat Agriculture Farms (Pvt.) Ltd.	4 Aug 83	43-Alnoor Building, Bank Square, Lahore	Corporate Agriculture Farming
4. Be Be Jan Agro Farms Ltd.	27 Feb 89	P-74, Liaqat Road, Faisalabad	Corporate Agriculture Farming
5. Gurmani Agro Farms and Livestock (Pvt.) Ltd.	5 May 90	43-Alnoor Building, Bank Square, Lahore	—
6. Prime Agriculture Farms (Pvt.) Ltd.	21 Apr 98	2nd Floor, Buland Markaz, 33, Blue area, Islamabad,	—
7. Millennium Farming (Pvt.) Ltd.	9 Jul 98	H # 218, St. 21, E-7, Islamabad	—
8. Pir Jhando Farms (Pvt.) Ltd.	19 Aug 02	Deh Shah Mureed, Tapo Sungal District, West Karachi	Corporate Agriculture Farming
9. Red Rose Corpn. Farming Co. (Pvt.) Ltd.	4 Oct 02	11/1/1, Model Colony, Malir, Karachi	Corporate Agriculture Farming
10. Corporate Farming Company (Pvt.) Ltd.	20 Jan 04	76, St. 23, G-10/1, Islamabad	Corporate Agriculture Farming
11. Sunrise Fields (Pvt.) Ltd.	11 Mar 04	2, St. 17, F-7/2, Islamabad	Corporate Agriculture Farming
12. Mcgill Farms (Pvt.) Ltd.	8 Apr 04	32-A, Garden Block, New Garden Town, Lahore	Corporate Agriculture Farming
13. White Gold Farms (Pvt.) Ltd.	29 Apr 04	Sandowali, Near Aroop More, Sialkot Road, Gujranwala	Corporate Agriculture Farming
14. Maan ki Dua Agriculture Pakistan Ltd.	26 May 04	Office #11, 1st Floor, Nadir House, I.I Chundrigar Road, Karachi	Corporate Agriculture Farming
15. Moinuddin Poultry (Pvt.) Ltd.	17 May 04	Shop # 53,Jamia Rashida Masjid, Malir Tanky, Malir Town, Karachi	Corporate Agriculture Farming
16. Royal Agri Farms (Pvt.) Ltd.	28 Jun 04	4-Bawa Park, Upper Mall, Lahore	Corporate Agriculture Farming
17. Royal Agri Products (Pvt.) Ltd.	28 Jun 04	4-Bawa Park, Upper Mall, Lahore	Corporate Agriculture Farming
18. Red Rose International (Pvt.) Ltd.	9 Jul 04	51-Faisal Block, Azam Gardan, Multan Road, Lahore	Corporate Agriculture Farming
19. FAHMS (Pvt.) Ltd.	20 Jul 04	104-B, New Muslim Town, Lahore	Corporate Agriculture Farming

ing is being pushed rapidly through the corridors of power for immediate implementation. This clearly indicates that there is lack of coordinated effort on the part of various ministries of the government. Similarly, genetic engineering is recognized as a risky technology worldwide which is also being promoted by our government.

NOTE

1. For summary of the evidence, see Martin Bouque and Ingrid Bekkers (1997), 'Silent Spring II? Recent Discoveries of New Threats of Pesticides to Our Health', *Food First.* Also refer to Sandra Steingraber, *Living Downstream: An Ecologist Looks at Cancer and the Environment*, New York: Addison-Wesley.

REFERENCES

Altieri, Miguel A. and Jean More von der Weid (2000), 'Prospects for Agro Ecologically Based Natural-Resource Management for Low-income Farmers in the 21st Century', *Global Forum on Agricultural Research Conference*, Dresden, 21–3 May.

Bayliss-Smith, T.P. (1984), 'Energy Flows and Agrarian Change in Karnataka: The Green Revolution at Micro-scale', in T.P. Bayliss-Smith and Sudhir Wanmali, *Understanding Green Revolutions: Agrarian Change and Development Planning in South Asia*, Cambridge: Cambridge University Press, pp. 167–70.

Bouque, Martin and Ingrid Bekkers (1997), 'Silent Spring II? Recent Discoveries of New Threats of Pesticides to Our Health' (http://www. agtrade policy. org/output/resource/SDTI.pdf#search=% 22Bouque% 2C%20 Martin %20and%20Ingrid). Also refer to S. Steingraber (1997), *Living Downstream: An Ecologist Looks at Cancer and the Environment*, New York: Addison-Wesley.

Brown, Lester (2000), 'Challenges of the New Century', *State of the World 2000*, New York: The Worldwatch Institute.

Hasnain, T. (1999), 'Pesticide Use and Its Impact on Crop Ecology: Issues and Options', *SDPI Working Paper 42*, Islamabad: SDPI.

Jeyarathum (1990), 'Acute Pesticide Poisoning: A Major Health Problem', *World Health Statistics Quarterly*, 43, no. 3, pp. 139–44.

Keeney, D. (1989), 'Towards a Sustainable Agriculture: Need for Clarification of Concepts and Terminology', *American Journal of Alternative Agriculture*, 4, pp. 101–5.

Lappe, F.M. et al. (1998), *World Hunger: Twelve Myths*, New York: Grove Press.

Mariano, R.V. (1996), 'Threatening Food Self-Sufficiency: GATT's Impact on the Grains Industry', *The GATT: Philippine Issues and Perspectives*, Quezon City: Philippine Peasant Institute.

McGinn, A.P. (2000), 'Phasing Out Persistent Organic Pollutants', in Lester Brown et al. (eds.), *State of the World 2000*, New York: Worldwatch Institute.

Natural Resource Defense Council (NRDC) (1996), *Summary of EPA Data*, May, vol. 61, no. 85.

Postel, Sandra (2000), 'Redesigning Irrigated Agriculture', *State of the World 2000*, New York: Worldwatch Institute.

Rosset, Peter (1996), 'Input Substitution: A Dangerous Trend in Sustainable Agriculture', working paper no. 4, International Council for Sustainable Agriculture, p. 4.

Rosset, Peter et al. (2000), 'Lessons from the Green Revolution: Do We Need New Technology to End Hunger?', *Tikkun Magazine*, vol. 15, no. 2, March–April, pp. 52–6.

Shiva, V. (1999), 'Why Industrial Agriculture Cannot Feed the World', paper presented at the International Forum on Agriculture Workshop, Cuenevaca, Mexico.

——— (1995), *Staying Alive: Women, Ecology and Development in India*, New Delhi: Kali for Women.

——— (1991), *The Violence of the Green Revolution: Third World Agriculture, Ecology and Politics*, Penang: Third World Network.

Smith, Carl (1988), 'Exporting Risk—US Hazardous Trade, 1995–6', *Pesticide News*, no. 40, June, pp. 4–5.

Wackernagel, Mathis and William Rees (1998), *Our Ecological Footprint: Reducing Human Impact on the Earth*, Gabriola BC and Philadelphia: New Society Publishers.

CHAPTER 8

Anjuman Mazarain Punjab: The Struggle Continues

Aasim Sajjad Akhtar

One of the most enduring legacies of the British rule in the subcontinent has been the structure of landownership in the rural areas. The traditional patron–client relations in rural areas have been altered and substituted by the imperial masters with laws that fit in their design. Such a change has damaged the centuries' old agrarian relations. The interest of the British was to do things they deemed proper and benefited them. But such changes have been detrimental to the interests of the small peasantry and helped land concentration in the Punjab in a few hands. A century ago canal-irrigated agriculture was introduced into west Punjab, what is now Pakistan, popularly known as the 'canal colonies'. The landed elite readily accepted this new form of agriculture. This wild area was cleared and turned into a cultivable region. Soon, it became the most fertile region in the whole of Punjab. For this purpose peasants from eastern Punjab were motivated to settle here. The loyalists of the Raj, soldiers and others who were to be honoured were allotted land here. The migration associated with the development of the canal colonies was an important part of the social change.

Thousands of east Punjabis were displaced and moved across to the western part of the province, including what are now the districts of Okara, Sahiwal, Khanewal and Sargodha, where they in turn displaced the settled population. These areas were largely forested, and the migrants were brought in to clear the land and develop the canal colonies. This was a massive social upheaval and brought about unparalleled demographic shifts in the Punjab. Ironically referred to by the British as 'colonists', the migrants were promised proprietary rights of land once it was arable. The land turned out to be the richest in the Punjab, and therefore, it was not surprising that the British retained control over it, rather than giving it up as promised to the 'colonists'.

One hundred odd years later, this land remains the most fertile in the province, and at least a part of it is now the most disputed. In Punjab and Sindh, *jagirdars* captured most of the irrigated land through bidding from the British. Some portions, however, were retained by the state. Approximately 68,000 acres of state-owned land in Punjab is now the site of the most compelling resistance movement that the country has witnessed in the past few decades. Spread out over 10 districts, this land is tilled by the families of the migrants from a century ago, now numbering almost one million. As sharecropping tenants, they have been surrendering harvest shares to the state since they settled in the areas. Over the past four and a half years since 2000 in quite a remarkable fashion, these tenants have engaged in a comprehensive and sustained display of civil disobedience by refusing to pay any form of tribute to the state while demanding ownership of the land itself.

Anjuman Mazarain Punjab (AMP) is the representative organization of the one million people residing on this land across the province. AMP is a radical people's movement reminiscent of an era not so long ago when radical land movements were not such a rarity. In this day and age, AMP's struggle is historic given the general apathy and depoliticization that characterizes Pakistani society. The movement itself, as such movements tend to be, was a rather spontaneous outcome of a dispute over the working relationship between the administration of Okara Military Farms—the largest of the many state farms spread out over 17,000 acres—and tenants. Subsequently, after General Pervez Musharraf had made repeated policy announcements that state land would be allotted to the landless, AMP developed a distinct identity as a movement for land.

The above-mentioned dispute emerged in June 2000 when the tenants of Okara Military Farms were told that they would no longer be considered tenants but instead would have to sign limited year contracts. Therefore, they would pay cash rents to the authorities and not a share of their harvests. The tenants refused, well aware that under existing tenancy laws, they were protected from eviction, but would not be if they became contract workers. As pressure increased on them to accede to the administration's demands, the tenants grew agitated, and soon started to organize large-scale public meetings to develop consensus on their outright rejection of the contract proposal.

Within months, the movement spread to Pirowal and Khanewal where another fledgling effort to organize tenants was taking shape. In Pirowal, the Punjab Seeds Corporation (PSC), a wing of the agriculture department, controls the land. Similarly, farms operated by maize and cotton research departments in Sahiwal, by the military in Lahore, Sargodha, and Multan, by rice research departments in Kala Shah Kaku and Faisalabad, by the livestock department in Sargodha, Sahiwal, and other departments in a number of districts were contacted.

It is impossible to attribute the emergence of such a movement to any one particular factor. While the contract dispute and the state's announcements on allotments acted as catalysts for the resistance to take shape, they do not account for anything beyond the initial reaction. The tenants' collective memory of displacement has definitely had an influence on them. It is natural for them to be extremely averse to the possible threat of another displacement. It is also true that there has been systematic abuse of power by local authorities on all farms, quite likely another source of resentment, and therefore resistance. Finally, the tenants of these farms have, at some level or the other, lobbied for ownership rights at different junctures in the past, sowing the seeds for a radical movement to evolve in the present.

From the very early stages, the AMP has been a mass movement. The regularity with which thousands of people have engaged in public action is astounding given the prevailing political environment. Soon after the movement started, the slogan *malki ya maut* (ownership or death) was adopted. The incredible rhetoric emanating from the government on land allotments to the landless acted as a tonic that changed the very character of AMP—it ensured that the tenants started articulating a vision instead of just reacting to a threat. The movement expanded rapidly. A couple of political parties sped to the scene, with the Jamaat-e-Islami's Kissan Board making itself the most prominent. The Kissan Board stayed with the movement through its early days but disappeared on one too many occasions when it was required to stand with the tenants through state repression. Left parties and organizations had arguably the most involvement with the movement, and remain the closest to the movement today. Some of the organizations worth mentioning are the Labour Party of Pakistan and the People's Rights Movement. The latter is a confederation of social movements working towards structural changes in the state of Pakistan while resisting the negative impacts of corporate globalization. AMP is its constituent member.

AMP's argument is very straightforward—if the government is giving state land away, then the tenants already working on them should be the first on the list. This is a legitimate claim in its own right. The argument becomes even more convincing when it becomes known that none of the agencies controlling the land that the tenants are tilling, including the military, actually have any legal right to it. The land in question is the property of the Government of Punjab. At one time or the other, the military, PSC, and other agencies, leased the land from the provincial government but these leases have long since expired.

As such, the agencies involved, including the military, have asserted that giving up the land in question is not even a debatable issue. The military has publicly stated that it owns the land, even though numerous documents highlight an exchange between the Ministry of Defence and the Punjab Board of Revenue (BOR), the latest in 2001, in which the BOR clearly refuses the former's request for the land to be permanently transferred to the military. Meanwhile, retired army officials who currently run the PSC have made laughable claims that the AMP is being secretly propped up by multinational companies. These companies are supposedly propagating their own agenda, thereby threatening the future of state-run agricultural research farms.

This rather opaque accusation glosses over the fact that one of the state's most celebrated policy initiatives at present is Corporate Agriculture Farming (CAF). This initiative will see unlimited tracts of state land handed over to agri-business companies on long-term leases. Capital intensive farming, it is said, will lead to unbridled progress in the agricultural sector. Unfortunately, there is very little mention of the fact that by its very nature, the initiative compromises the land reform agenda in the country, and is likely to have a devastating impact on small and landless farmers, thereby contributing further to the already unsustainable push toward large cities. It is in fact true that there are plans to eliminate land ceilings that were instituted in the 1977 reforms by Zulfikar Ali Bhutto.

The argument that farms run by agricultural departments for research purposes are being threatened also runs hollow. First, many of the farms have a significant area in self-cultivation alongside the area that is cultivated by the tenants. Second, many farms have been operating in massive deficits for decades. There is significant evidence that harvest shares collected are not deposited in government accounts, and are instead deposited into private pockets. The tenants have also made clear that even after a transfer of land, they would be more than willing to continue cooperating with the research departments by experimenting with seeds that they are given—this is the only real form of research that takes place on the farms at present in any case.

Finally, it should also be known that the PSC has been slated for privatization for some time now. In Pirowal and on all the other farms, there is every reason to believe that the state's gung-ho attempts to make the tenants sign contracts are related to the CAF initiative. These farms represent the only possible option for companies wishing to come in and run their businesses on already fertile land with readily available contract workers. Incidentally, and quite conveniently, these contract workers will not be granted basic rights given to labour in the industrial sector, such as the right to organize and assemble.

Against this backdrop, the state has also engaged in routine and unprecedented violence to intimidate the tenants into silence. In January 2002, Bashir Ahmed was shot and killed on the orders of the Deputy Director, Renala State Military Farms, a serving colonel in the Pakistan army. In May 2002, two tenants were killed in Dipalpur and Pirowal, again on the orders of either serving or retired army officers. Subsequently in June 2002, a man and woman in Pirowal died after facing continuous siege from the police for weeks. In August 2002, a tenant in Okara was killed after being tortured to death while in police custody.

Okara Military Farms—and the 1,50,000 people who live in 19 villages there—were besieged for over a month prior to the death of Suleiman Masih in August 2002 by the police and the Rangers. The siege ended only when the tenants were forced to sign (or put their thumbprints on) the contracts while a gun was held to their heads. A number of observers, including foreign journalists, and even a retired head of the Navy, verified reports that coercion was being used to make the tenants sign the contracts. Relatives of tenants had been issued show-cause notices from their state employers' warning of disciplinary action if they did not 'convince' the tenants to cooperate. Some of these employees were subsequently fired.

Another siege took place in May 2003, the culmination of which was the death of another tenant, shot by the Rangers. While tension in Okara and other farms has lessened in recent times, harassment is still commonplace. A local journalist, Sarwar Mujahid of *Nawa-i-Waqt*, has been arrested twice for supporting the tenants cause. He is being held under a non-bailable anti-state case. Meanwhile many AMP activists remain in jail, some without charge, some carrying fairly serious injuries. Hundreds more face criminal cases.

Nonetheless, the Director General of Pakistan Rangers and other high-level army generals have been adamant that the tenants were ready and willing to cooperate with the authorities and that it was a handful of troublemakers, including outside parties, that were inciting the otherwise peaceful tenants. He also suggested that these outside influences had links to the Research and Analysis Wing (RAW), the Indian intelligence agency. The outrageous propaganda is not surprising, especially in light of the fact that the AMP has seriously shaken the establishment. The same Director General of Rangers, who has a rank of Major General, has publicly warned the landed elite to be wary of the reaction of tenants tilling private lands if the military was forced to confer proprietary rights on the agitating tenants in Punjab.

Such statements, and the general attitude of the military toward AMP, highlight just how important land is to the military itself. It would not be inaccurate to suggest that the military is amongst the largest landholders in the country, if not the largest. It is clearly the military's persistent efforts to usurp land through institutionalized means that has allowed the landed elite to maintain the kind of political influence that it has. The military needs friendly political forces (as the 2004 elections clearly indicate), and the landed elite needs support to compensate for its eroding power base in the rural areas. With the alliance consummated, it is no small wonder that the already marginalized groups such as the rural landless get the short end of the stick.

In the first instance, it more or less amounts to heresy for sharecropping tenants to revolt against their existing tenure relationship (just or unjust is besides the point) with the state—and that too on land controlled by the military. Second, it is probably beyond the military's conception that the revolt has persisted for as long as it has, in spite of all of the harassment that has taken place. The fact that all of this is happening in Punjab is also problematic. Traditionally, similar uprisings have been witnessed in other provinces and the state has moved strongly to crush them. In Punjab, the visibility and the risk are higher for the establishment, as is the actual political significance of the struggle. Finally, it is inconceivable for the military to actually succumb to what is effectively the democratic voice of the residents of the area, because if they do so, the Pakistanis will finally realize that it is possible to defy the prevailing pecking order.

While the state may feel it has at least temporarily dealt with the issue by forcing tenants to sign the contracts, the fact remains that even on farms other than Okara, tenants remain mobilized and radical. The most resistance actually took place in Pirowal in June/July 2002 when electricity, water, and phone connections were severed and the standing cotton crop was destroyed. Despite this, the tenants withstood the pressure and refused to sign the contracts. This consciousness persists. Meanwhile, the authorities have not been able to bring a halt to the spate of public meetings and actions in the area, even if they have contained them to a certain extent for now.

The AMP's demand is for the Punjab government to be the ultimate arbitrator in the entire affair. The AMP has consistently asked to meet with representatives of the Punjab government, saying that they will not sign any agreement with any party that is not the owner of the land in question. Furthermore, they say—and quite rightly—that it is impossible to have meaningful dialogue with any party given the kind of repression that has been let loose on them. Their struggle is important for a number of reasons. It has genuinely alerted activists and intellectuals, as well as the flag-bearers of the *status quo*, that, however, weak it may be, a culture of resistance still persists in Pakistan.

In a movement that has defied all existing social and political norms, the role of women perhaps stands out as the most extraordinary characteristic of all. The picture of a mobilized *Thappa* (a small wooden bat used to wash clothes) Force comprising women has generated much support for the movement. *Thappa* Force has actively faced aggression from the authorities in the form of batons, tear gas shells and even guns. Women have been the most vocal critics of the military and other agencies responsible for the repression. They have begun the process of redefining their role in the societies within which this social movement is embedded. While it will require much more to be done to rectify the immense gender disparity that still persists in Okara or Khanewal, there should be no doubt that oppressive social relations within society are often fundamentally altered through struggle.

The desperate struggle of women in these areas also highlights how the threat of eviction is the most acute for them. In Pakistan, there can be no doubt that displacement affects women most, in large part due to restrictions on women in public life. It should also be remembered that displacement remains an accepted consequence of 'development' in the country. Whether it is the rural or urban landless facing eviction, or those affected by mega projects, the trauma of displacement remains a marginal concern to policy makers. This conception has not changed from the time that the canal colonies were built, and reveals how the state continues to cling to obsolete and elitist notions of progress.

The role of the Christian minority in the movement represents another attempt to challenge these elitist notions. A very significant percentage of the tenants are Christian, in some areas the figure is as high as 40 per cent. It is this Christian minority that has been the most radical and aggressive in the struggle, despite a history of being the most oppressed community in the country. The state has engaged in considerable propaganda to dispel the impression that any resistance remains, but the fact is that the anger and resentment on all farms is possibly greater than ever before.

The AMP has recently linked up with a variety of other peoples' movements under a larger umbrella called the People's Rights Movement. As much as at any other time in recent history, politicized elements of society have responded to the AMP's calls for support. The media in particular has been an invaluable asset, and in recent time, influential international newspapers such as the *New York Times, Washington Post* and the *Guardian* have also supported the tenants' cause. International agencies such as the New York-based Human Rights Watch (HRW) have documented and condemned the atrocities committed by the military on the farms. (The report is available on their website.) That said, and the scale of the movement being what it is, there is much more that supporters could yet do. There is a stark need for this movement, and the multitude of issues that it encompasses, to become a much more significant part of our political discourse. Unfortunately, as with most other issues that truly impact society's ethics and attitudes, the level of debate and discussion is very limited.

The movement has exposed the shortcomings of our political culture, while emphasizing that there are still spaces available to articulate a vision that challenges a state that continues to operate on the same principles that characterized British rule. It is important to realize the symbolism of this movement at a time when society tends more and more toward parochialism. It is also important to see how the state has responded to the AMP's struggle in the context of the history of canal colonization, and the concentration of land in rural areas. There remain many important lessons to be learnt, and hopefully many battles for landless tenants to win.

The role of the [illegible] ... [illegible] to challenge [illegible] ... emergence of the [illegible] ... that what has become the most radical and aggressive [illegible] ... of the most oppressed [illegible] ... [illegible]

[illegible] ... supported [illegible] causes [illegible] international agencies such as the [illegible] have documented and [illegible] ... The report is available [illegible] ... and the scale of [illegible] ... that support [illegible] ... much more significant part in our political discourse [illegible] ... with [illegible] that [illegible] ... [illegible]

The movement has [illegible] ... on the same [illegible] ... time when [illegible] ... see how [illegible] has responded to the [illegible] struggle in the context of the history of [illegible] colonization, and the concentration of land in rural areas. There are many important lessons to be learnt and hopefully many battles for landless groups to win.

CHAPTER 9

Punjab Peasantry and the Green Revolution: An Activist's Perspective*

Balkar Singh Dakaunda and Pashaura Singh Sidhupur

Seventy per cent of India's population is dependent on agriculture, yet the country's agricultural development is far from satisfactory. There are only a few small regions besides Punjab and Haryana that have developed agriculture. On the whole it may be said that only a small part of India is agriculturally developed of which Punjab is in the forefront.

Why did Punjab develop? This development took place because the country faced a serious food grain crisis in 1952. The United States had then already developed into an economic power. US-based multinational companies and other institutions like the Ford Foundation prepared an agriculture policy in the wake of the Indian crisis that served to promote the market interests of these companies. This policy was given the name of Green Revolution and helped the companies gain 'easy profit' through the sale of agricultural inputs like farm-related machinery, insecticides, irrigation projects and seeds to this new market which had been created. The rich peasantry and to some extent the common peasants too were lured to produce for the market by the subsidies extended to them for agricultural inputs, easy loans and support prices. The pro-imperialist Indian government publicized this as an important step towards 'self-reliance'.

The above-mentioned processes linked the agricultural production of this region with the complex system of market exploit-

*This paper originally written in Punjabi is translated by Birinder Pal Singh.

ation. Agricultural production did rise, and it is also true that the living standards of the common peasantry improved. Business, transport, services, electricity, education and health facilities also developed though in a limited way. These sectors also absorbed some labour rendered surplus in the agricultural sector. This also led to the development of urbanization to an extent. Those who got engaged in the above services improved their living standards somewhat more than the common peasantry.

The peasantry did get some short-term gains from the development of the Green Revolution, but the major chunk of the profit was garnered by the national and international industrialists, multinational corporations, merchants and usurer-commission agents sitting in the market.

SOME EYE-OPENERS

- The Green Revolution that started in the 1960s increased the consumption of fertilizers from 21 thousand tonnes in 1970–1 to 15 lakh tonnes in 2003–4. The prices of chemical fertilizers also registered an enormous increase of 150 per cent during this period.
- The price of diesel was 71 paise per litre while wheat was 76 paise per kg in 1967. Today wheat sells at Rs. 8.50 per kg and diesel sells at Rs. 32 per litre.
- The productivity of land under Green Revolution has started declining over the last two decades, but the cost of production is rising. The central government and the Commission for Agricultural Costs and Prices are not only shirking from fixing a small rise in the Minimum Support Price (MSP) but are planning to introduce a policy that may help them avoid fixing the minimum purchase price (*ghato-ghat khareed kimat*) and leave the farmers at the mercy of the open market.
- Subsidies have been withdrawn from all quarters. The Congress government in Punjab withdrew the subsidy (2003) since it was supplying free electricity to the farmers. In the wake of privatization/corporatization of the Punjab State Electricity

Board and the ensuing crisis in the production of electricity, the possibilities of it becoming costlier and its supply becoming more erratic have increased.

- Agriculture today in the regions of 'Green Revolution' has become increasingly capital intensive. It has led to an enormous rise in agricultural expenses due to expensive machinery, and ever rising costs of fertilizers, insecticides, diesel and so on.

THE ISSUE OF WATER

The issue of water is very significant for the Punjab peasantry. Agriculture is dependent on water. If there is no water, there is no agriculture. Practical experience tells us that modern agriculture is essentially dependent on tube wells rather than canals. According to government statistics, 96 per cent of the total area under agriculture in Punjab is already under irrigation. But sadly enough, in 1970–1 the area under canal irrigation was 1,292 thousand hectares and 1,591 thousand hectares under tube well irrigation. The corresponding figures for 2002–3 stand at 1,148 and 2,828 thousand hectares (*Statistical Abstract of Punjab*, 2003). But according to our estimates the figures for canal irrigation are only on paper as in fact it is only one-third of the total irrigated area since canal water is not released for days together. Water does not reach the 'tail' regions of the farms and the waterways for canal irrigation have been dis-mantled by the farmers, thus necessitating their dependence on the tube wells.

According to a survey of the Government of India, 80 blocks out of a total of 118 blocks in Punjab have the hard clay soil. The ground water is falling at the rate of 30 cm per year. The farmers are forced to bear the expenses of installing deeper submersible pumps as the government and the agricultural university are least concerned about alternative solutions.

The need of the hour is to procure more canal water and implement rational crop diversification, but more important is to find solutions beneficial to the peasantry.

GREEN REVOLUTION HAS TURNED PALE

The agricultural production of this region is not only stagnant but actually on the decline. According to a survey of the Punjab Social Sciences and the Health Forum:

> The yield of crops per hectare that had become stagnant earlier has now registered a decline. The yield of wheat has declined from 4,696 kg per hectare in 1999 to 4,563 kg per hectare in 2000–1. The yield of rice has also registered a similar rate of decline.

The Forum reports further,

> The Punjab peasant who has liberated India from the disgraceful agreements of the like PL-480 and made the country self-reliant in food production is now itself standing at the verge of destruction. The rate of growth of agriculture is now less than the all India average. The Punjab peasantry is badly caught in the debt net.

All avenues to join other services have been virtually closed to the Punjab population dependent upon the agricultural sector. The land has undergone division after division and the machinery has increased manifold over the years. This has been followed with a constant decrease in agricultural production subsequently leading to a decline in the income of the peasantry. That is why the debts on peasantry are increasing.

Of the total debt on the peasants, the commission agents account for 60 per cent share and the remaining 40 per cent is from commercial banks and cooperative societies. All these agencies charge high rates of interest. It is so high that about 70 per cent of the small and middle peasantry cannot return even the short-term agricultural loans advanced to them. All 'transactions' are made on paper only. As a matter of fact this debt is increasing each day as interest is added to the principal amount as also the legal and illegal administrative expenses. As on today, each small and middle peasant is carrying a debt of Rs. 12,000–15,000 per acre or about Rs. 30,000–38,000 per hectare. About 80 per cent of the Punjab peasantry is presently under serious economic crisis. Of the 11 lakh Punjab peasant families, 6 lakh are below the poverty line while two-thirds population of the state is directly or indirectly engaged in agriculture.

The small and to some extent the middle peasants are openly or stealthily mortgaging their lands while others have started selling it. It is unfortunate that the peasantry that considers the earth as 'mother' has to do this. Hundreds of peasants have committed suicides due to the poor economic condition of the previous years.

UNEMPLOYMENT AND SEMI-UNEMPLOYMENT

Another and more serious and vicious aspect of the agriculture sector is that unemployment is rising at a fast rate in Punjab. This problem is widely spread in both rural and urban areas of the state. According to the official statistics there are about 15 lakh unemployed persons out of which 11 lakh are in the rural areas (Government of Punjab, 1999). Many noted economists have argued that at least half the labour force engaged in Punjab agriculture is surplus and must be withdrawn. As a matter of fact there is no work left in the Punjab villages. The central government has stated on oath before the Commission for the National Review of the Working of the Constitution, that 'the means and opportunities for creating rural employment have reached their limits'.

These aspects of Punjab agriculture had become manifest, though in lesser proportion, during the 1980s and even before that. The Khalistan movement in Punjab from 1980 to 1992 was also a consequence of stagnation and economic deterioration. There could be numerous differences on the analysis of this problem but no one can deny this fact. Punjabis have borne the brunt of this problem. Thousands of people have fallen prey to individual or state terrorism. We must work hard to check the explosive situation of the day going the wrong way.

STOCKPILES OF GRAIN AND THE PUBLIC DISTRIBUTION SYSTEM

- Of the 31 Indian states and union territories 18 had identified their respective populations below the poverty line till 1 January 2001, but most of them did not make available ration cards to the concerned persons thereby forcing them to buy ration at the market rates.

- There is also contradiction in the figures of those identified as poor. The government has identified 15 crore people below the poverty line while the Planning Commission records more than double, that is 26 crore.
- The Public Distribution System (PDS) had a stock of 8 million tonnes in 1998–9 that had reduced by 70 per cent in the first quarter of 2000.
- The Food Corporation of India (FCI) had 64 million tonnes of wheat and rice in its stock in 2001. The government decided to release 5 million tonnes of wheat for sale in the open market. Interestingly, the government sold wheat to the poor people through the PDS at the rate of Rs. 8.50 per kg and to the businessmen at the rate of Rs. 6.50 per kg.

The essence of the argument is that the government is shying away from giving food to the hungry and the half-fed people of the country. On the other hand, the food grains stockpiled in government godowns are getting spoiled thus alluding to the fact that Indian food stock is 'overflowing' which is why the government is not in a position to buy fresh stock from the farmers. This is a clear indication of the government's apathy towards the poor people and the hardworking peasantry.

All these things are being enforced under the name of New Economic Policy (NEP), at the instance of the WTO for the globalization of the economy. But, in our understanding this is suicidal for the working classes, especially the small peasantry. This aspect is elucidated below with reasonable arguments and suitable data.

WTO'S ILLUSIVE POLICIES AND ARGUMENTS

The first argument that is given by the WTO in favour of globalization is that doing away with the subsidies to the peasants and removing the barriers to agricultural trade between the countries will create a 'free world market' where farmers can obtain remunerative prices for their commodities. A section of the rich farmers in Punjab is going beyond this point in arguing that with the abolition of subsidies in the developed countries, the costs of production of their farmers shall become relatively higher compared to us, hence we shall be able to gain substantially from the world market. This argument is nothing more than an illusion since the Punjab peasant is unable to take his crop to commission agent of his choice as 80 per cent peasants are compelled to sell their present crop immediately before the next sowing begins, no matter what the price. Only a miniscule section of the rich peasantry can afford to store the crop such that it could be sold at their convenience when the prices are high. How can the Punjab peasant compete with the MNCs owning 2 lakh acres of land? The small peasant cannot even compete with such surplus producers of food grains as in the US or Australia in his home market. The US has listed India among those 13 countries that have been identified for aggressive export promotions of edibles, fruits, vegetables, milk and its products. It is not true that the developed countries will stop subsidies to their farmers. As a matter of fact they are not only continuing the subsidies but increasing them using the loopholes in the policies of the WTO. According to the US Secretary of Agriculture, the subsidies will not be reduced, rather these would be increased to $600 million over the next five years.

Another argument that is being furthered is that the peasant should come out of the wheat/rice/cotton cycle and instead go for export-oriented commercial crops so that he can earn rich profits in the world market. Former Chief Minister Prakash Singh Badal and now (2005) Amarinder Singh have given a hawker's cry to the debt-trapped peasantry to try cropping fruits and flowers. The diversification of crops and contract farming are illusions since there is neither a permanent and stable market in this age of privatization nor are there any support prices or subsidies for such ventures. These are only hollow suggestions. For instance, in the year 2004–5 *basmati* rice was subjected to contract farming but neither the contracting agencies nor the government bothered about its purchase. The peasants could hardly sell it at Rs. 1,000–1,200 per quintal in the open market, half the price compared to Rs. 2,400 per quintal the previous year. This is the state of affairs of crop diversification and contract farming.

Secondly, the problem with export-oriented agriculture is that it would be directly linked to the world market whose ups and downs have already ruined the countries of Latin America and Africa. It happened so in Latin America because the farmers were motivated to grow only bananas. No doubt these countries did benefit from it initially, but as the banana prices crashed in the world market their producers could not be saved from ruin.

Thirdly, if the export does take place the important question is who is going to benefit from it? Are the producers of the crops going to get the benefit? Never at all. Let us have a look at the domestic market itself. Whenever the prices of onion or potato rise, only a section of the businessmen reap the benefit or at best the top strata of the rich peasantry who work as agents of the MNCs.

Fourthly, the issue of agriculture-dependent industry is being talked about. We should be clear that in the private sector the industry with modern technology would be controlled by the MNCs. Only such companies shall reap the profit through modern food processing rather than the peasants who produce tomato for the Pepsi.[1]

Thus, publicity about the above policies and projects is meant to confuse the peasantry. It is a delusion. It is rather clear from the points made below that the WTO is like a hungry wolf lying in wait:

1. Our peasantry is being asked to reduce the production of food grain and switch over to commercial crops; the government agencies are dilly-dallying from entering the market and not fixing the minimum support price. Such a move would push the peasantry further into poverty. It can also push the country towards a serious food crisis. India will be forced once again to tread the path that led to the shameful compromises under PL-480.
2. State corporations like the Dairy Development Board, MARKFED and Punjab Agro Industries are seeking alliance with private monopoly houses and the MNCs for the establishment of nodal points. This would only benefit the corporations to the detriment of the public and the cooperative sector.
3. The new seeds and patent laws would not only enhance the poverty of the peasants but also raise other serious problems.

For instance, the MNCs would get the new seeds patented and subject these to research modifications of their choice with the sole purpose of making profit. It is the considered opinion of many pro-people social and health scientists that such genetically modified seeds will be harmful for the health of the people. But the MNCs like Cargill and Monsanto have already strengthened their monopolist grip at the international level.

Two Fresh Examples

Over the last five years the producers of cotton/*narma* (American cotton or *Gossypium religiosum*) are worst affected by the American bullworm. Production, however, has improved this year (2004). India today has a stock of 22.7 million bales of cotton whereas the consumption requirement is only for 17.7 million bales. The country imported 3.5 million bales from Australia at the time when the Punjab peasant brought his cotton and *narma* to the market. This led to the crashing of its price to Rs. 1,600–1,700 per quintal from an earlier high of Rs. 2,500–2,600 per quintal. Our peasant organization Bharatiya Kisan Union (BKU [Ekta]) and many others protested against the import but the Union and Punjab governments failed to take any action.

The case of coconut oil is no different. Its price came down to Rs. 2,825 per quintal from an earlier rate of Rs. 6,825 per quintal in 1997 as a result of the import of Palm oil.

NEED OF THE HOUR: ADVANCE WITH DETERMINATION

We not only feel but have firm conviction that the WTO does not consider the peasantry, the village poor and other working classes as parts of the Indian society. This is clear from the living conditions of these classes as on today and what these are going to be tomorrow.

The peasant unions must strengthen themselves organizationally to pull out the Punjab peasantry from the present crises. All unions must launch struggles with determination not only for their demands but also for the strengthening of their respective organizations. Their demands should be:

1. Struggle continuously for the 'Government of India to come out of the WTO'. The government must be pressurized to stop immediately the implementation especially of agriculture-related agreements.
2. The government should be forced to develop truly cooperative agriculture for the poor, small and middle peasantry. Its *modus operandi* should be worked out in consultation with peasant organizations.
3. Large industries should be established in the state of Punjab in the public sector, and private sector and the MNCs must not be allowed in this venture. This industry should mainly be related to agriculture and it must be employment-oriented so that unemployment and semi-unemployment can be eliminated.
4. The allied occupations that should be developed too must be employment-oriented. For the development of allied occupations like poultry, and dairy, long-term loans and subsidies should be granted liberally to the peasants.
5. The loans of the poor and small peasants should be waived. Henceforth, these classes should be given interest-free loans. The middle peasants should be charged nominal interest.
6. The peasants should be given free crop insurance schemes.
7. The agricultural produce must get appropriate price and its purchase should always be guaranteed.

Our organization, the BKU (Ekta), will have to deal with the government tough-handedly even for the partial acceptance of the above-mentioned demands. The union and the state governments are ignoring the peasant struggles at the moment. The Punjab peasantry must organize struggles against the government in the light of its glorious past, since the future is full of challenges as the living conditions of the poor and small peasantry is deteriorating day after day. These classes are confronting poverty and the threat of a dark future. The middle peasantry too has overcome its confusion and is now steering on the course of struggle for its bright future.

It is now clear that leaving aside a small section of the capitalist landlords and rich peasants, the rest of peasantry is fighting against imperialism, besides its complaints against the WTO.

Only if the peasantry is mobilized for a struggle can it obtain a few concessions and relaxations. In the recent past our organization has struggled against the moneylending commission agents and the New Economic Policy and succeeded in retrieving a section of the poor and middle peasantry from hopelessness. An indisputable evidence of this fact is that the suicide rates of the peasants have decreased substantially in those areas where our organization has been active.

NOTE

1. The reference is to the Pepsi plant in Patiala district. Before the establishment of the plant the peasants were promised employment on a preferential basis. They were also asked to grow tomatoes that Pepsi would buy *in toto*. Once the plant started working, neither of the two promises were kept by the Pepsi.

CHAPTER 10

Viability Crisis of the Small and Marginal Farmers and Response of the Farmers' Movement in Punjab

Sucha Singh Gill

In the decade of the 1990s Punjab's agriculture nearly stagnated after the potentials of Green Revolution technology were exhausted. This is also reflected in the declining returns from agriculture to farmers in the wake of relatively fast rising costs of production compared to rise in the prices of agricultural produce. The system of farming in the state is purely based on family enterprises of the farmers who are divided into different classes of operational landholdings. The farmers with better resource base cultivate large landholdings while those with poor resource base cultivate small holdings. The emerging crisis in the system of farming has affected all the farmers but has hit very hard the weakest among them. The poor classes of the farmers have come under severe pressure and are fighting for their sustenance. Although the farmers' movement has been active for a fairly long time, and the farmers have organized a series of mobilizations, some of these being very effective too, yet there are no signs of any relief to the poor classes of the farmers. This paper addresses to the issues of viability of the marginal and small farmers and the response of the movement to their problems.

VIABILITY CRISIS

The Punjab Agricultural University (PAU), Ludhiana, organized a two-day brainstorming meeting on *Farmers and Farming in Punjab* on 23–4 October 1998. This meeting was attended by the leading economists and policy makers of the country. Among the participants was Dr. Manmohan Singh, the prime minister of India. The base paper prepared by the PAU economists brought out that after meeting all paid out costs, an average 2 hectare farm in 1995–6 gave Rs. 40,740 as income with wheat-paddy rotation, and Rs. 42,180 with wheat-paddy-dairy farming. This income included rent for land, returns on the fixed capital and family labour. The average per capita income in the state being Rs. 16,044 which worked out to Rs. 96,264 for a six-member family, was higher than (more than double) the income yielded by a 2-hectare farm. It was further stated that for achieving the state-level per capita income for a family of six members the ownership of land required was estimated to be 4.56 hectares. This meant that 75 per cent of the farming families were below the average per capita income of the state. Raising productivity by even 10–15 per cent from the present level, as 75 per cent potential has already been achieved, would not take small and marginal farmers anywhere near the average per capita income in the state. The paper concluded that

> the economic condition of the farmers owning 2 hectares of land is unsatisfactory and the economic condition of the farmers owning less than one hectare is worst and cannot be improved with existing technology and cropping system. (PAU, 1998)

The participants in this brainstorming session agreed to this observation. And while expressing serious concerns on the situation, made several recommendations to the state government that remain largely unimplemented till date.

Along with the academic and official recognition of the poor plight of the small and marginal farmers, two related developments are worth mentioning. One is related to the entire village community falling under heavy debt burden in the dry cotton belt of southern Punjab in the districts of Bathinda, Mansa and Muktsar. Two villages (Harkishanpura in Bathinda district and Shergarh in Muktsar) have been put on sale. The village panchayat of Harkishanpura has put a signboard outside the village with writings in bold letters in Punjabi: Village for Sale. The Government of Punjab ordered investigation in this matter and found that the land of the farmers was mortgaged mostly to the commission agents. The same is also true of Shergarh village in Muktsar district. The southern districts of Punjab mainly depend on canal irrigation as the sub-soil water is saline and tube well irrigation is not effective. The scarcity of water is adversely affecting cultivation and the declining water table in tube well-irrigated areas is burdening farmers with increasing investments in the frequent deep boring of tube wells.

The second development is concerned with the phenomenon of farmers committing suicide in the state. Though there has been relatively high concentration of this unfortunate phenomenon in the southern districts of Sangrur, Mansa and Bathinda, yet such cases have also been reported from other districts of Punjab (Bhalla et al., 1998). Detailed investigations into this phenomenon bring out that along with other factors, economic hardships and the pressure of creditors for recovery of loans have been forcing farmers to commit suicide. It is further brought out that the majority of the victims of suicides were marginal and small farmers who had fallen neck deep into debt due to crop failures or heavy expenditures incurred on medical treatment or social ceremonies, especially the marriage of daughters or sisters (Gill et al., 2000; Iyer and Manick, 2002).

The non-viability of small and marginal farmers with existing cropping pattern, technology and farming system is pushing them into heavy indebtedness due to crop failure or on expenditures arising out of medical or social requirements. Though many of them are able to sustain, yet several of them are breaking down under economic pressure and are forced to leave cultivation. The less fortunate among them, when threatened and unable to find alternative acceptable avenues of livelihood, act violently against both themselves and others. Statistical support to this observation is also found from the data on operational landholdings in the state. The data reveal that in 1970–1 combined share of marginal and small operational holdings in the total operational holdings was 57.54 per cent which declined to 38.62 per cent in 1980–1, increased to 44.63 per cent in 1990–1 but further declined to 29.67 per cent in 2000–1. In absolute terms the decline in the marginal holdings was 1,72,908 hectares (from 2,95,668 to 1,22,760) and in the small holdings by 30,772 hectares during 1990–1 and 2000–1. This is contrary to the rising number and proportion of semi-medium, medium and large landholdings. The large decline in marginal and small holdings has resulted in decline in the number of total operational holdings by 1,19,578 hectares (Table 10.1). The decline in the number and proportion of small and marginal holdings and rise in the number and proportion of semi-medium, medium and large holdings has led to shifting of land cultivation under relatively larger holdings, leading to concentration of land. The process of growing land concentration with relatively bigger holdings was stalled during the 1980s but was restored after 1990–1 when normalcy was restored in the state after the turbulent years of Sikh militancy, especially after 1992–3. The operation of lease market under reverse tenancy, the declining water table and deteriorating quality of soil also work against the poor farmers. They cannot afford to meet the cost of deepening of tube wells and it is difficult for them to maintain costly implements like tractors, trollies, pump sets, threshers, reapers, and so on.

The present organization of cultivation, that is, individual cultivation, does not ensure their viability. It has been brought out by a well-researched study that such small and marginal farmers who depend on cultivation alone cannot sustain and survive in the present form as their expenditure is more than their income. Among such farmers who are able to diversify into dairying, poultry, vegetable growing, etc., or one of their family members is able to find a job in the public or private organized sector are able to survive. The survival of small and marginal farmers in Punjab requires increase in farm size, enhanced productivity, rise in income from allied and off-farm activities and drastic reduction in fixed investment, crop expenditure and consumption expenditure (Singh, 2006). Present-day Punjab agriculture is highly capital intensive. The individual cultivation mode has added to overcapitalization of agriculture. This has added to fast increase in the cost of cultivation where fixed costs have been rising very fast since the 1980s. It is estimated that an average utilization of tractor is 20 per cent of its capacity in Punjab. On small farms utilization of tractor is as low as 4 and 16 per cent on medium farms. On the large farms, the utilization is definitely higher. The same is the case with other implements such as threshers and diesel engines. This makes small

TABLE 10.1: NUMBER OF PROPORTION OF OPERATIONAL HOLDINGS BY SIZE GROUPS

Size groups (in hectare)	Year					
	1970–1	1980–1	1985–6	1990–1	1995–6	2000–1
Marginal (0–1)	5,17,568	1,97,323	2,56,259	2,95,668	2,03,896	1,22,760
	(38.63)	(19.21)	(23.61)	(26.48)	(18.65)	(12.32)
Small (1–2)	2,60,083	1,99,368	2,07,639	2,03,842	1,83,453	1,73,070
	(18.91)	(19.41)	(19.08)	(18.25)	(16.74)	(17.25)
Semi-medium (2–4)	2,81,103	2,87,423	2,90,837	2,88,788	3,20,340	3,28,230
	(20.44)	(27.99)	(26.72)	(25.85)	(29.31)	(32.91)
Medium (4–10)	2,47,755	2,69,072	2,54,142	2,61,481	3,05,794	3,00,174
	(18.08)	(26.20)	(23.90)	(23.41)	(27.98)	(30.70)
Large (10 and above)	68,883	73,941	70,553	67,172	79,610	72,360
	(5.00)	(7.19)	(6.75)	(6.01)	(7.28)	(7.26)
Total	13,75,392	10,27,127	10,88,430	11,16,951	10,93,073	9,97,370
	(100.00)	(100.00)	(100.00)	(100.00)	(100.00)	(100.00)

Note: Figures in the brackets are in percentage of the total.

Source: *Statistical Abstract of Punjab 2002*, Government of Punjab, Chandigarh, 2004. Figures for 2000–1 are collected from the office of Statistician, Department of Agriculture, Government of Punjab, Chandigarh.

and marginal farms non-viable. The practice of resorting to custom hiring creates problem of timely availability of machinery for various agricultural operations. The present production system works to the disadvantage of the small and marginal farmers due to their inability to reap economies of the scale unlike medium and large farmers. Compared to other categories of farmers, the small and marginal farmers face fast rise in the costs of cultivation when other things remain the same. This reduces their net income per acre. The falling incomes make difficult for them to replace old implements. The falling water table is adding further to their woes. It is estimated that during the next ten years (2001–11), the deepening of the existing bores and the installation of submersible pumps will cost the Punjab farmers almost Rs. 2,000 crore (Sidhu and Johl, 2002). The per acre outstanding debt on small and marginal holdings was Rs. 10,105 against Rs. 4,228 for medium holdings and Rs. 4,230 for large holdings in 1997–8 (Shergill, 1998). This puts additional pressure on the marginal and small farmers in the form of interest payments. With increasing exposure to the market forces, the economy of the small and marginal farmers is coming under increasing pressure, making it unviable. These farmers with poor resource base face further difficulties due to shrinking employment in the agriculture sector, virtually no recruitment in the public sector, and very little employment in the organized private sector. They have no skills other than agriculture for an alternate living. The production system does not favour them either, as they are being crowded out of the land lease market. The technology also is not useful to them and the ecological changes are adding further to their woes. Therefore, the small and marginal farmers find little space in the agriculture of the state. This class of farmers needs support both from the state government and the farmers' organizations more than any other class of farmers.

RESPONSE OF THE FARMERS' MOVEMENT

In contrast to other parts of India, farmers' movements in Punjab have been active for the last couple of decades. Earlier the Kisan Sabha movement organized both agricultural labourers as well as the cultivators in an effort to build worker–peasant alliance. But soon it was realized that they have to be organized separately under different banners. In opposition to the left dominated Kisan Sabhas, the farmers' movement began to be organized independent of party politics in the state. This movement arrived on the scene with a big bang towards the close of 1970s and emerged as a big force under the banner of Bharatiya Kisan Union (BKU). From the very beginning the BKU in Punjab has organized farmers on the issues of input prices and prices of agricultural produce. Except for the earlier experimentation to develop agro-processing units under farmers' cooperatives, and fighting corruption in the government departments dealing with the farming community, the BKU mobilizations have concentrated on the prices of agricultural outputs, market clearance of the agricultural produce and prices of agricultural inputs, especially issues of subsidy. This remains the dominant agenda of the farmers' movement in the state. The Punjab Kisan Sabhas and other organizations of the left also continue to follow this agenda.

Mobilizations are done to pressurize the state and the union governments to meet the farmers' demands and shift government policies in favour of the farm sector. During the last decade and a half after the militancy phase ended in Punjab, the farmers have undergone several phases of mobilizations but mostly on issue of market clearance, procurement prices of wheat and paddy, clearance of arrears of sugar cane sold to the sugar mills and restoration of free electricity for tube wells.

The issue of indebtedness of the farmers has been raised several times. This issue surfaces repeatedly, especially whenever a farmer commits suicide. A suicide by a rich farmer, Kuljit Singh of Sakraudi village in Sangrur district in 1998 following the brutal killing of other family members, attracted a large number of farmers and their leaders to his *bhog* (final ritual at death) ceremony. On the sad occasion, all the farmers' union leaders declared that they would launch a movement for the abolition of debt of the farming community and designated it as the 'debt liberation movement'. But nothing of this sort happened. Occasionally, the farmers' organizations have been able to block the auction of land of indebted farmers when creditors are either commission agents or cooperative/commercial banks. But they have never raised the issue of dispossession of the poor peasants from land by the rich/large farmers through mortgage, sale or via the lease market. This is not on the agenda of the farmers' organizations. In fact, this has not become a part of the consciousness of the farmers. The agricultural experts and leaders of farmers' organizations agree in personal discussions that the present cropping pattern, farm practices and mode of individual cultivation or organization of agriculture cannot save the small and marginal farmers from liquidation within the present regime of the market. In spite of this, the farmers' organizations are not raising the issue of transformation of society or larger social issues such as crippling of public funded rural education and health or the employment of rural youth in the wake of sharp declining employment in agriculture. They have forgotten the agenda of rural development. During the 1980s it was talked about in the farmers' rallies that 40 per cent of the budget of the government be spent on rural development. It is no longer talked about now. At the same time, the reform agenda such as working on farms/work culture, reduction of unnecessary expenditure on social ceremonies connected with marriage, birth or death and festivals or spending on liquor and drugs are rarely figuring in the agenda of the farmers' organization. Similarly, gender discrimination practised in such a crude form as female infanticide is not a part of the farmers' movement. Such issues are left for the government or for other organizations, although they affect the farming community directly.

The farmers' organizations are engaged in mobilizing farmers for raising their issues and put up a struggle in the form of demonstrations and rallies. But they have yet not involved themselves in experimenting and developing alternatives in production, marketing or agro-processing. In the rest of the country experimentations are going on in farmers' cooperatives, organic farming or integration of farm production with agro-processing, and with success too. But no such lead is being taken by the farmers' organizations in Punjab, though there is a lot of scope. To use Shankar Guha Niyogi's phrase, 'they are for struggles not for construction'. Thus, they are happy to provide temporary relief on a few issues rather than work for the liberation of poor farmers or facilitate the smooth transition of rural society.

LIMITED INTERVENTION OF FARMERS' ORGANIZATIONS

The farmers' movement in Punjab has a glorious past. Earlier under the leadership of the All India Kisan Sabha, the farmers spearheaded a successful movement against the Betterment Levy in 1960s. During the 1970s, the BKU undertook several successful movements on the issues of input and output prices. The movements achieved a grand success when almost the entire rural Punjab was under its influence during 1980–4. This culminated into a successful siege of the Punjab Raj Bhawan (governor's residence) at Chandigarh from 12 to 18 March 1984, in which 50,000–60,000 farmers participated. They ran their own community kitchen at the site. The farmers from the nearby villages maintained a regular supply of food, fuel, milk, etc., for their fighting bretheren. The agitation was massive but peaceful and attracted support not only from within the state but also from the neighbouring state of Haryana. Following this success in terms of demands conceded and the organizational gains, the BKU-led movements attracted a lot of media attention. The Union also took over the entire rural Punjab and banned entry of the officials of the Punjab State Electricity Board, the Cooperative Department, the Revenue Staff and the Police into the villages without the permission of the BKU. The Union also established 'jails' in the villages to detain corrupt officials of the various departments of the Government of Punjab or those who attempted repression and harassment of the farmers (Gill and Singhal, 1984). The BKU also planned to stop the movement of grains from Punjab to other states but its mobilization process collapsed when the Government of India placed entire Punjab and Chandigarh under the control of the army and imposed curfew on 3 June 1984 during Operation Blue Star, meant to flush out militants from the Harmandir Sahib at Amritsar and numerous other gurdwaras all over the state. Subsequently, when the entire democratic process was paralysed, the BKU continued the process of mass mobilization (Gill, 2000).

Mobilization under the BKU was massive because it was declared to be free from political parties. This point of strength turned into weakness when the movement came to have massive sway over the farmers. The political parties began to woo the movement leaders and co-opt them in important positions. This led to a split in the movement in 1989 when Bhupinder Singh Mann, the all India president of the BKU was nominated a member of the Rajya Sabha (upper house of the Parliament) by the V.P. Singh-led government. The faction of the BKU led by Ajmer Singh Lakhowal subsequently got involved in Akali Dal politics, causing further split in 1994 when its third faction, the BKU (Ekta) was born from BKU (Lakhowal). The third faction of the BKU was led by Pashaura Singh Sidhupur.

The Sidhupur faction was supported by a large number of leftists. The BKU (Ekta) was further split into two, one led by Pashaura Singh Sidhupur and the other by Joginder Singh Ugrahan in 2003. In 2004, BKU (Krantikari) separated from BKU led by Pashaura Singh Sidhupur. Another faction led by Ruldu Singh Mansa formed the Punjab Kisan Union in 2006. The last split came in December 2006 when BKU (Ekta) was formed by Balkar Singh Dakaunda out of BKU (Ekta) of Pashaura Singh Sidhupur. Similarly, in 2006, the Rajewal and Mann factions formed their separate BKUs. Thus, the original BKU at present is split into eight organizations where, with the exception of one, all are carrying the tag of BKU. In addition to these eight organizations there are three Kisan Sabhas, one Kirti Kisan Union and one Kisan Sangharsh Committee. Thus, the farmers' movement is split into thirteen organizations all claiming to represent the farmers in the state. The split in the organizations and fragmentation of the movement has generated a lot of bitterness and rivalry among the different factions. Considerable amount of energy of the leaders and cadres is wasted in justifying the split, thus making the movement weak. At the moment, each of the farmers' organizations has a small geographical area (one or two districts) of operation for mobilization. Each of the organizations individually is not in a position to put sufficient pressure on the state government for getting its demands conceded. Attempts to build a united front of these organizations to protect the farmers' interests have often been tried but all of these could not be brought under one umbrella. At the most, two fronts, one consisting of six organizations and another consisting of four organizations, came into existence while the Lakhowal faction and the factions led by Bhupinder Singh Mann and Balbir Singh Rajewal remained outside the two coordination committees. Operationally the activities of these coordination committees remained symbolic in nature and failed to revive effectively the mobilization of the farmers at the state or the national level. Moreover, split and fragmentation of movement provides enough space for unscrupulous elements to avoid discipline and provide them an opportunity to indulge in the violation of rules of the organization or join the rival factions when tuned out of one organization for indiscipline. This makes both the organizations and the movement physically and morally very weak.

In the wake of fragmentation of the organizations and decline in the morals of the established leadership, the movement has become intellectually weak. There are several issues which are crucial for the very survival of the farmers but their organizations are not thinking about these. One crucial issue is related to the degradation of soil and fast declining water table threatening the sustainability of agriculture. Another is related to the shrinking quantum of common property resources including village ponds and panchayat or *shamlat* land. The farmers' movements had an agenda of social transformation during the 1960s, 1970s and 1980s. But this has been abandoned since the 1990s. Contrary to the expert advice that paddy transplantation should not be done before 15 June, the farmers start transplantation process in the first week of May itself, leading to massive pumping out of the sub-soil water. Though agricultural experts are crying to save the subsoil water, the farmers' organizations have been silent spectators for the last many years. Similarly, one does not hear anything from the farmers' organizations on the use of genetically modified (GM) crops and organic farming. The collapse of rural education and rural health systems have stopped the process of vertical mobility, especially from agriculture to secondary and tertiary sectors, but these are no longer issues of struggle for these organizations.

The various factions of BKU and other farmers' organization have been raising issue of indebtedness but have not launched any movement over it. This is at a time when the business and industry have got one-time settlement relief to the tune of Rs. 1,86,000 crore of loans from nationalized commercial banks. In most of the cases the loan payment has been 25 to 50 per cent of the total. The movement remains trapped in the input (especially free electricity for agriculture and free water from canals) and output prices and the market clearance of wheat and paddy. The lack of capacity for sustained movement to free (especially poor) farmers from the debt trap and procurement of any relief for the families of farmers committing suicides have alienated the small and marginal farmers from the movement. The movement is now largely confined to the middle and rich farmers.

The first decade of the twenty-first century has posed a fresh issue of development and displacement in the wake of the Government of India's policy of establishment of Special Economic Zones (SEZ). This involves acquisition of agricultural (farmers') land up to 5,000 hectares at one place for the development of a SEZ, development of infrastructural facilities followed by several concessions involving subsidy of several thousand crores of rupees to businesses. The farmers' movement is opposed to the acquisition of land for SEZs but is unable to articulate the aspirations of farmers whose land is being acquired. When it comes to clinch an issue for individual farmers, it boils down to the rate of compensation to them. But other occupational categories such as agricultural labourers and petty traders who are also displaced from there disappear from the agenda of the movement. Even for small and marginal farmers who lose the only source of their livelihood, the issue of rehabilitation and employment are of paramount importance. The farmers' movement is stuck either in total opposition to SEZs or in the rate of compensation at market prices that however remains undecided due to extremely low rate at which sale-purchase of land is registered by the farmers to save stamp duty. This has particularly come out in the case of farmers' resistance to land acquisition in three villages (Fathepur Chhanna, Dhaula and Sanghera) in the Sangrur district during 2006–7.

To sum up, it is clear that the farmers' movement is fragmented and weak. It has lost its past glory and vision. Intellectually and organizationally the movement is weak compared to the issues confronting the farmers, especially the small and the marginal. Being dominated by the rich farmers, the movement fails to take up the issues of poor farmers and other rural poor in the state. Consequently, the land is shifting from the farmers to the non-farmers, that is, the commission agents, industrialists and businessmen and also from the poor farmers to the well-off farmers. The viability crisis of the small and marginal farmers is pushing them out of agriculture. The farmers who do not have skills other than cultivation are therefore hard pressed and under lot of pressure to leave cultivation and enter an uncertain future. The farmers' movement is under the influence and hegemony of medium and large farmers who pay little attention to the specific needs and protection of the poor farmers. These organizations take up the general issues of farmers against commission agents, banks, the government and the corporate sector but when the viability of the small farmers is threatened these organizations have no programme.

REFERENCES

Bhalla, G.S. et al. (1998), *Suicides in Rural Punjab*, Chandigarh: IDC.

Gill, S.S. (2000), 'Agrarian Change and Farmers' Movement in Punjab: A Case Study of BKU', in Harish K. Puri and Paramjit S. Judge (eds.), *Social and Political Movements*, Jaipur: Rawat Publications.

Gill, S.S. et al. (2000), *Suicides in Rural Punjab* (in Punjabi), Ludhiana: Association for Democratic Rights, Punjab.

Gill, S.S. and K.C. Singhal (1984), 'Punjab: Farmers Agitation: Response to Development Crisis of Agriculture', *Economic and Political Weekly*, vol. XIX, no. 40.

Government of Punjab (2002 and 2004), *Statistical Abstract of Pubjab*, Chandigarh: Pubjab Government.

Iyer, G.K. and Mehar Singh Manick (2002), *Indebtedness, Impoverishment and Suicides in Rural Punjab*, Delhi: Indian Publishers and Distributors.

Punjab Agricultural University (1998), *Farmers and Farming in Punjab*, Ludhiana: Proceedings of Brainstorm Meeting.

Shergill, H.S. (1998), *Rural Credit and Indebtedness in Punjab*, Chandigarh: IDC.

Sidhu, R.S. and S.S. Johl (2002), 'Three Decades of Intensive Agriculture in Punjab: Socio-Economic and Environment Consequences', in S.S. Johl and S.K. Ray (eds.), *Future of Punjab Agriculture*, Chandigarh: Centre for Rural Research and Industrial Development.

Singh, Mandeep (2006), 'An Economic Analysis of Small and Marginal Farmers in Punjab', unpublished Ph.D. dissertation, Department of Economics, Ludhiana: PAU.

CHAPTER 11

Sikh Militant Violence: A Manifestation of the Economic Crisis in Punjab

Birinder Pal Singh

A unique feature of the landmass called Punjab has been that its people have always been engaged in an active interaction with external forces since antiquity. Niharranjan Ray puts it straight: 'History therefore taught the Panjab and her people one very important lesson, namely, not to forget or be oblivious of temporal or secular situations of any given time or space, howsoever engrossed one might find oneself in matters of the mind and the spirit' (Ray, 1975: 105). I have already explained in the introduction to this volume that the Sikh religion and philosophy further concretized this bond between the two situations.

After India's Independence in 1947, there had been strikes and agitations—*morchas* in Akali parlance—for one reason or another, largely against the central government. First of all there was a struggle for Punjabi Suba to be delineated on linguistic lines as other states in India. After a long drawn struggle for more than a decade the demand was fulfilled. Then came the issue of implementation of Punjabi as an official language of the state besides obtaining the Punjabi-speaking areas left out to the neighbouring states. The demand for possessing Chandigarh, the 'city beautiful' and the capital ciy, is yet not settled. During the Emergency years of 1975–7, the Akali Dal had been in the forefront of the country-wide agitation launched for the restoration of democracy and fundamental rights of the citizens. True to the character of the Akali Dal all these agitations or *morchas* had been non-violent and peaceful but for certain minor aberrations here and there.

After the Emergency, the Akali Dal passed a resolution in 1978, called the *Anandpur Sahib Resolution*, an earlier draft of which was prepared in 1973. It demanded a federal structure of Indian government and administration such that the states or provinces could enjoy greater autonomy in their socio-cultural and economic development.[1] As a part of this struggle for greater federal autonomy, the Kapoori Canal Morcha was launched to stop linking the Sutlej and Yamuna rivers meant to help the neighbouring state of Haryana to have a share in the Punjab waters. It is during this *morcha* that various factions of the Akali Dal and the militant leader Jarnail Singh Bhindranwale came together. Since then, until June 1984, the Akali Dal launched a number of *morchas* like blocking roads (*rasta roko*) and stopping the trains (*rail roko*) for the implementation of the *Anandpur Sahib Resolution.* More and more problems were emerging and the situation was becoming ever more complex, not due to the reluctance on the part of the agitators but due to the lackadaisical attitude of the central government and the politics of the then ruling Congress party.

A brief introduction to the Akali Dal is now necessary (Grewal, 1996; Singh, 1978). The Shiromani Akali Dal (SAD) was formed in December 1920 to protect the interests of the Sikhs. It had been actively engaged in anti-colonial struggle at various levels and kept the nationalist spirit of freedom alive. The SAD contested the 1937 elections and bagged 10 out of the 14 contested seats. After Independence in 1947, it prospered as a regional political party that gradually developed roots in Haryana and Delhi too. In Punjab it remains a major party along with the Indian National Congress (INC). Traditionally, the latter had an urban and the former a rural base besides the fact that Akali Dal was predominantly Sikh dominated whereas the INC had a mixed following with a slant towards Hindu dominance. Thus, the following of the Akali Dal was largely constituted by the peasantry. It had and still has a weak urban base. For the Akali Dal politics, the geographical territory of Punjab, its language and culture are virtually synonymous with Sikhs and their religion which is why it is often dubbed by the opposition as communal. One feature that characterizes this party is its non-violent approach to agitation.

Their non-violence is remarkable since the Punjabi society and more so its rural component subscribes to a value system that valourizes power, dominance and violence. Both before and after Indian Independence, a violent stream of agitation/protest had been accompanying its non-violent counterpart. In the latter phase over the last few decades we have witnessed two such movements, primarily based on violence and both involving the youth predominantly from the rural areas. The first was the Naxalite movement inspired by the ultra-left and Marxist–Leninist philosophy in the late 1960s. These youth subscribed to the philosophy of annihilation of the bourgeoisie, the landlord and the agents of the state since that alone could liberate the poor and the landless. Though the active movement was crushed with a heavy hand in a short span, its legacy was carried forward by numerous Marxist–Leninist groups, some of whose members are still underground. The Punjab Students Union is a recognized radical union which is still active amongst the students in the state.

Another, still more violent movement took off in 1978 and traumatized the state and the nation equally for more than a decade. It was the Sikh militancy spearheaded by the chief of a Sikh seminary, Damdami Taksal, Jarnail Singh Bhindranwale following an incident of killing of thirteen Sikhs at Amritsar on Baisakhi day (13 April) allegedly by the Nirankaris when a *jatha* went to their camp to lodge a protest against violating the Sikh principles. This incident snowballed into more violence, more killings both of Sikh and the Hindu public men—political leaders, administrators and policemen. The educated unemployed Punjabi youth was fascinated with the slogan of changing the socio-political system, and fighting in the name of *Panth* for a religious cause.

Militant groups mushroomed in different parts of the state carrying out operations in their own area and each proving to be more committed to the *Panth* than the other.[2] The Government of India had to launch Operation Blue Star to capture Bhindranwale, his confidantes and other militants from Harmandar Sahib, the Golden Temple. Indira Gandhi the then prime minister, was held responsible for this act of sacrilege and was assassinated by her Sikh security guards within less than six months. It led to three days of massacring of Sikhs in Delhi and other cities all over the country. A brief lull prevailed in Punjab before it could break into an intensive and extensive violence all over the state. This phase continued till the end of 1992 when militancy came to an abrupt end (Singh, 2002).

This militant movement has been characterized differently by various people and political parties on the basis of their ideology. The Indian government led by the Indian National Congress (I) labelled it a 'separatist' (separation from India), 'disintegrationist' (breaking the integrity of the Indian nation), 'fundamentalist' (a la Khomeini of Iran), and a 'terrorist' movement. The then dominant party in opposition, namely the Bharatiya Janata Party called it an 'anti-Hindu' and an 'anti-national' movement interested in creating Khalistan, a Sikh theocratic state. They consider India a nation of the Hindus, and Sikhism a sect of Hinduism. The Communist parties characterized militancy as 'extremist', 'undemocratic', 'fascist', 'obscurantist', 'ethnic' and a 'fundamentalist' movement.

But the SAD and the militant groups in particular spearheading the movement did not believe in such characterizations. They considered themselves 'fighting for a just cause' that meant their rights, against discrimination of the Sikhs and for their freedom of belief, expression and action. The militant groups were crying aloud that they had a 'distinct goal', a 'clear self-perception', a 'professed logic of violence' in their movement (Singh, 1997). Each one of these issues is a subject of an independent inquiry. In the present paper an attempt has been made to cull out from the statements of the militants the reasons that pertain to the socio-economic conditions of the state having a bearing on their movement. No doubt the militants came from all sorts of castes, classes and religions yet the small peasantry dominated the leadership and its cadre.[3] Obviously, they articulated their material demands though cloaked in religious terminology.

The militants claimed to establish an independent sovereign state of Khalistan where 'the Sikhs could experience the glow of their freedom', a remark made by Jawaharlal Nehru in April 1946.

Each one of these groups professed to have a clear vision about the social, political, cultural and religious aspects of the future society of their dreams. But in the present paper the focus is only on the socio-economic issues raised by them in their proclamations—handouts, posters, booklets and press releases.[4]

A perusal of their statements below would show that the militants were articulating such issues of poverty, inequality, exploitation, capitalism, the class character of the Indian ruling class, the limitations of the Indian Constitution, rights of individuals and communities, problems of development, and so on that afflict the Indian polity and society in general and the Punjabi society in particular. The focus of their concern was economic exploitation and political domination by the central government. The militants not only raised the issues that were specific to Sikhs but other marginal and oppressed communities in India as well. They not only highlighted those issues specific to Punjab but also to other states. Besides, they also identified the classes and castes that could be their allies in their liberation struggle.

Singh Khalsa proclaim that by

> Khalsa *raj* is meant a country (*desh*) or rule of pure people (*shudh lokan*). Khalsa *raj* would be free from evil (*paap*) and evildoers. Pain and suffering would not be there in theory or practice. It would also be farther from economic and cultural exploitation (*lutt-khassutt*). The Khalsa *raj* would be a truly democratic state. (Singh Khalsa, 1985: 1)

They made it clear that their struggle was not directed against the poor of any caste or religion. 'It is desirable that we must preach amongst the poorest of the poor and the lower castes, and among those who have been misled by the ideology of the Brahmin and the Bania' (ibid.: 33). It is suggested that 'the Hindu theory of Karma has made people timid, cowards and lazy. It has rendered them incapable of understanding what kind of spring (*bahar*) could economic and social change bring in their lives' (ibid.: 15). They argue that both Brahmin and Bania are the rulers of the day. These reactionary forces are out to destroy the Khalsa since they do not tolerate the rise of any revolutionary philosophy. In such a situation:

The Khalsa must win over the poor, and people of lower castes to his side who are slaves, both economically and socially. They must be made aware that the ideologies of Brahmin and Bania (*Brahmanwad ate Baniawad*) are the cause of their ignorance (*jahalat*), poverty and weakness. (ibid.: 17)

Finally, they impressed upon the lower classes and castes the need to realize that it is only Khalsa who could uplift them and ensure complete independence (*sampooran azadi*).

The Panthic Committee, a powerful apex body of five militant outfits headed by Dr. Sohan Singh, a former Director of Health, Punjab government, also suggests that in Khalistan:

Lack of education and social backwardness will not be allowed to be an obstacle in their way. Nor the monopoly of education will be allowed, as a tool, to snatch the rights of the illiterate as the children of the rich, and urban residents leave behind the rural and poor children. The rich enjoy the boons (*nihmatan*) of nature much more than what is due to them while the children of the rural and the poor remain victims of illiteracy, poverty, diseases and backwardness generation after generation. (PC[PM] 1986: 26)

Sarbat Khalsa, a congregation of the entire Khalsa, was held at the Akal Takht on 26 January 1987. It adopted a *Gurmata* (a resolution adopted by all those gathered in the presence of the sacred Sikh scripture, the *Guru Granth Sahib*) that reaffirms the self-perception of each militant outfit:

This congregation of today proclaims for the information of the whole world that the Khalsa, who wishes the welfare of all (*sarbat da bhala*), shall never attack the poor and the oppressed (*mazloom*). The present struggle of the Sikhs is directed against those plundering and destructive raiders (*lotuan, dharwian ate vinashkarian*) who have assaulted our principles (*sidhant*), gurdwaras (*gurdham*), *Guru Granth* (*Bani*), our form/dress (*bana*) and truthful earnings (*kirat kamai*). . . . This assembly of Sarbat Khalsa strongly endorses the armed struggles of the peoples of the world, especially those in India who are fighting against the tyrant colonial rule (*zalam samraj*) for their rights and independence . . . to maintain their cultural existence (*sabhiacharak hond*) and nationality. This congregation recommends the formation of all religious minorities front . . . to confront the Delhi government (*Dilli sarkar*). (*Gurmata*, 1987: 3)

The militants were not only crying for their freedom and vocalizing their problems but also took notice of the living conditions of the poor. They had identified agencies and institutions responsible for their poverty. Singh Khalsa stated categorically: 'Our struggle is against anti-Khalsa powers (*taktan*) like the big Bania (*vadde-vadde banian*), capitalists, feudal lords, big (*vadde-vadde*) Brahmins [*sic*] and official informers' (Singh Khalsa, 1985: 32).

The Babbar Khalsa International (BKI), the most dreaded militant outfit in the last phase (1989–92), also championed the cause of the poor who were getting poorer while the rich were growing richer. The poor peasant, however, gets exploited twice, initially at the time of selling his products and later while procuring essential commodities from the market. The BKI asserts:

> The Hindu capitalism [*sic*] intends to squeeze the poor economically, like a lemon, to such an extent that they could think of nothing more than mere subsistence and keep begging at their doors. These capitalists are leading a luxurious life after having sacrificed (*bali le le ke*) the means of subsistence, the sons, youth, honour and dignity of the poor. (Babbar: 14)

Wassan Singh Zaffarwal of the Khalistan Commando Force (KCF) also makes it clear in his interview that there could be no compromise with the Brahmins because they sit idle. The government of Khalistan will be based on the principles of Sikhism enshrined in the *Bani*. He says forcefully: 'We will not create a society where one human being is poor and sleeps on the road while his neighbour sleeps in the palace. We shall eliminate all remaining feudal and monopolist forces' (Pettigrew, 1995: 154). Another leader of the KCF declares: 'You cannot take money from any poor person, ever. We're clear on that. However, we shall impose a tax of the Khalistan government on the wealthy. We won't be forcing money out of them. We shall tax them' (ibid.: 163).

Singh Khalsa also holds similar views about the chief enemies of the poor people and the Khalsa:

> The feudal lords, Brahmins and the moneylenders (*soodkhor*) are supported by the Hindu colonialism [*sic*]. They are filling their coffers by sucking the blood (*khoon choos-choos ke*) of people whom they have made their slaves (*gulam*) both economically and politically. (Singh Khalsa, 1985: 45)

The responsibility is fixed on the exploiting classes 'who are befooling and deluding us through caste, religion and social hierarchy (*ooch-neech*) to serve their own interests' (ibid.). That is why it is the duty of the Khalsa to acquire political power since no philosophy can live long without it (ibid.: 10).

> Without this political sovereignty both the Khalsa *Panth* and the Sikh religion would be swallowed by the ideologies of the ruling elite like the feudal lords, Brahmins and Banias. And this sovereignty of the Khalsa cannot be obtained without an armed struggle. (ibid.: 11)

The inimical attitude of the militants towards the systems of domination and exploitation of the poor have also been stated in the *Document for the Declaration of Khalistan*. The Panthic Committee (Punj Membari) (PC[PM]) announced:

> The Khalistan government would like to distribute the natural boons (*kudrati nihmatan*) and meet the bare minimum needs (*zaroori zarurtan*) on humane basis (*manukhi adharan*). Monopolist and capitalist tendencies will not be allowed to influence the government machinery and peoples' thinking (*sochni*). (PC[PM], 1986: 26)

The Sikh Students Federation (SSF) also declared vociferously that:

> The Sikh struggle is directed only against those blood-sucking leeches (*lahu peenian jokan*), wicked (*dushtan*), tyrants (*jarwanian*), sinners (*papian*), and destructive (*vinashkari*) raiders (*dharwian*) who have made fatal assaults on their *Bani, bana*, gurdwaras (*gurdhaman*), culture and truthful earnings (*kirat kamai*). (Sikh Students Federation [SSF] 1989: 27)

The Federation then specifies the enemy:

> Therefore, the main targets of the Sikh struggle (*jaddo-jehad*) at the present moment are the Brahminic forces who have captured the Indian state. But this also includes those forces (*taktan*) in Punjab who collaborate with the central government (*kendari hakman*) for their economic interests and oppose the Sikh movement due to their political kinship (*siasi natedari*) with them. All such people, irrespective of their caste and religion are also included in the enemy camp and they would be dealt with accordingly. (ibid.: 27)

The identification of the enemy alone was not enough. The allies too were identified. Who could be their supporters? It has been mentioned above that the Panthic Committee specifically invited *Dalits* among other classes and minorities. Singh Khalsa also noted that certain castes and classes opposed the Khalsa right from its birth,

> there are also such forces who have always helped the Khalsa. These forces include the peasants, workers, lower castes, middle classes and other dominated and oppressed people. These forces will help the Khalsa in future since the Khalsa has itself emerged from the oppressed people. (Singh Khalsa, 1985: 44)

The BKI questions if the Sikhs today have closed their eyes to the tyranny of the capitalists and decided to lead a life of comfort and luxury. How could such people be called Sikhs? Because, a Sikh is one who dies fighting for justice and protects the honour of the poor (Babbar). This organization (BKI) doubts the possibility of a socialist revolution under the leadership of the 'so-called socialist revolutionary parties' and surmises if these were genuinely treading the socialist path. It is believed that true association of people (*sanjhiwalta*) cannot be raised on the 'foundations of atheism and hatred since it lacks sympathy for humankind. It could only be based on the pristine, social and spiritual principles of *Sri Dashmesh*' (ibid.: 17). 'It was a result of this association only that the Khalsa, who emerged from each backward class (*pachhri shreni*) could overthrow the well entrenched Mughal empire established over centuries' (ibid.: 15).

The SSF is not only apprehensive of external threats to the Sikhs from the Indian state and Hindu communalism, but is equally aware of those opportunist Sikh leaders 'who either belong to the rich class or ally with them. They are not remotely related to the Sikh principles and its way of life by way of their socio-economic status (*samajak arthak rutbe*) and lifestyle' (SSF, 1989: 24). These so-called leaders are in league with the capitalist class for their economic greed and political interests which have gone deep into their blood. Such Sikhs are responsible for the weaknesses and internal decay in the movement. This aspect becomes more glaring when it is compared with the earlier periods in the Sikh history. The Federation argues: 'So long the Sikh movement was guarded by the poor and the oppressed forces, it remained in perfect health (*naun-bar-naun*) and high spirits (*chardi kala*) with respect to its aims, objectives and principles' (ibid.: 24).

The Panthic Committee (PC[PM]) also proclaimed that this struggle of the Khalsa is not directed against any religion, community or caste. Its targets are those forces of evil that have chained Sikhs to slavery through force, deceit and cleverness. 'These Brahminic rulers (*hakam*) of Delhi who have been practicing treachery (*dagha*), deceit (*fareb*), tyranny (*zulam)* and force against the Sikhs for the last 44 years are the foremost enemies of the Khalsa *Panth*' (Panjwar, 1991). This statement identifies three types of enemies: (i) all those people in the security forces and bureaucracy who appear to be Singh but are, in fact, subservient to the Brahmin, (ii) all those political leaders or workers who have Sikh appearance but are mentally in league with the Delhi rulers, and in practice too take sides with them, (iii) the Sikhs who are not only morally corrupt and degenerated but also indulge in criminal and anti-*Panthic* activities (ibid.).

The Operation Blue Star (Indian Army's action inside the Golden Temple at Amritsar in June 1984 to flush out militants) and the massacre of Sikhs in Delhi in November 1984 (following the assassination of then prime minister, Indira Gandhi) and elsewhere have removed the veil from the face of the 'tyrant and the killer communal Hindus':

> Now it is not enough to identify the tyrants and the killers, but they should be eliminated following the Guru's command. The Khalsa has been created only *to destroy the tyrant* and the tyranny, and *for the protection of the poor.* (Babbar: 17) (emphasis in original)

Therefore,

> O Khalsa adorn yourself with weapons following the command of the Tenth Father (*Dashmesh pita*) and bring the present (Sikh/Akali) leadership on the right track to fight against tyranny. And if they refuse to oblige then O'Khalsa remove these obstacles on your way to liberation. (ibid.: 21)

Khalsa *raj* or Khalistan will never be served on a platter. Thus, the battle is inevitable following the dictum '*Kou kisi ko raj na de hai, jo le hai nij bal se le hai*' (No one offers a kingdom on a platter, it has to be annexed).

BKI is optimistic that such battles will usher Sikhs towards greener pastures:

The sacrifices of the martyrs will not go waste. . . . We *must make such a country/nation (desh) where the Khalsa is supreme, which has its own constitution, flag (nishan) and Nanakshahi currency and where we can enforce the principleof 'Welfare of all' following the principles of the gurus. In such a country the religious people, the poor and the workers could be protected from the exploitation of the tyrant and cruel capitalists and monopolists, so that they may lead a happy life of self-respect with dignity and honour.* (Emphasis in original) (ibid.: 22)

Once again in 1991, this organization reiterated its stand for the creation of a 'new society':

A new era is about to begin on the land of Khalistan. This new milieu will have exhaustive debates on the Khalsa culture, Khalsa vision, Khalsa rule and Khalsa society which will help us construct a beautiful model for the economic, political and social structural aspects of Khalistan. (BKI, 1991: 19)

But the BKI and all other militant groups are apprehensive about the realization of their vision of a 'new society' within the framework of the Indian Constitution. The BKI holds that it is the biggest hurdle in the creation of Khalistan. 'There is no place for Khalistan in this cartload of papers' (ibid.: 19). The Sikhs have already waited too long since 1947. There is no alternative but to reject this Constitution is the BKI's stand. It is cautioned that 'If we start any struggle without rejecting the Constitution, our struggle is bound to lose direction' (ibid.: 19). But what kind of a struggle: an armed struggle which must have harmony and coordination with the peoples' struggle for the establishment of an independent and sovereign Khalistan. It is further suggested that the given moment is most suitable to launch their struggle. 'The international situation is in our favour and India too is a victim of serious economic and political crises' (ibid.: 19).

The KCF and the Khalistan Liberation Force are also cognizant of the international situation and advise their sister organizations to understand the changes that are taking place throughout the world. 'It is so very necessary today as it had never been before' (Singh and Bhudhsinghwala, 1989: 26). It is in view of such developments that these organizations are bound to take their struggle to the international level:

It is clear from India's intervention in the Sri Lanka's Tamil problem that one sovereign state could intervene in the affairs of another sovereign state. Therefore, we would be wholly justified in accepting assistance from some foreign country'. (ibid.: 26)

The four militant organizations—Khalistan Commando Force (Panjwar), Khalistan Liberation Force (Budhsinghwala), Bhindranwale Tiger Force of Khalistan (Chhandran) and the Sikh Students Federation (Bittu)—also made a strong and fervent appeal at the Anandpur Sahib Convention in September 1991 to reject the Constitution of India which was referred to as a 'thief's mother' (*chor di ma*) and the 'root of all problems'. 'This heap of garbage looks nice on the Brahmin's shoulders only' (Panjwar, 1991: 3). They also stressed on the suitability of that moment to launch a direct action which of course was to be undertaken as a last resort. They suggested:

First of all we must remember that this battle is being fought on our own land. Therefore, we will exhaust all channels of the diplomatic world so that the war could be avoided. But we will not digress an inch from the path of obtaining an independent and sovereign Khalistan. (ibid.: 6)

These organizations also impressed upon the urgency of direct action:

The international situation is so congenial, splendid (*shandar*) and appropriate (*dhukwin*) that if the Khalsa now failed to shape its diplomacy to these conditions or failed to avail of the contradictions of the world (*sansar dian virodhtaian*) [*sic*] in its favour, then we must understand that we have ourselves prolonged the period of our distress (*khuari di miad*). (ibid.: 3)

These outfits also cautioned the Government of India that if tyranny against the Sikhs continued then the country would meet the same fate as Russia, once a superpower. They also issued a warning to all the countries of the world, the International Monetary Fund (IMF), and other international financial institutions that 'they must sign loan agreements with India on this understanding that *the people of* Khalistan *will not be a party to their repayment.* Because, not a fraction of these loans has been invested on the land of Khalistan. And we do not need it either' (ibid.: 5) (emphasis in original). On the contrary:

> The brave farmers of *our country* Khalistan, are feeding the empty stomachs (*bhukhe dhidd*) of crores (one crore = 10 million) of *Hindustanis* (we are not). These countless *Bhai* Ghanaiyas[5] in the service of humankind will maintain this tradition even after the recognition of Khalistan by the Indian government.
> (Panjwar, 1991: 5)

The SSF was not swayed by its religious affiliation alone in articulating the interests of Sikhs and Punjab. It was equally concerned about other communities. It was fully conversant with the socio-political situation of India. It argues that due to the communal behaviour (*vihar*) of the Indian rulers, the people have been 'compelled to launch struggles in one form or another for their economic, political, social and cultural independence' (SSF, 1989: 23). The SSF named all those Indian states where such struggles for liberation have taken varied forms to fight against

> the imperialistic exploitation and neocolonial suppressive rule (*samrajvadi lutt-khassutt te navbastiana damankari raj*) of the Indian rulers. At certain places especially in Nagaland and Tripura, the people have already taken to an armed struggle for their freedom as a result of state terrorism that is continuing there for several decades. (ibid.)

The SSF is also concerned with the status of the cultural and linguistic minorities and other nationalities, the tribals and Dalits who are being exploited economically by the capitalists and both politically and culturally by the ruling class. They liken India to a 'prison' from which all such exploited and oppressed minorities would like to escape. All such classes are interested in establishing a political system in which they could realize their economic, political, social, religious and cultural aspirations without any intervention from outside. 'Thus, the Sikhs are inclined to extend support and hand of friendship towards those struggling classes' (ibid.: 27). It is argued that the above-mentioned minorities could obtain democratic rights only

> If the present Centre-oriented aggressive state administration (*kendar-mukhi dhakkar raj parbandh*) is forcefully uprooted and a new federal structure is raised which ensures democracy and complete self-determination to the states in the true sense of the term. (ibid.: 27)

The militants were also critical of the processes of development, and of state's intervention and hence discrimination against Punjab, and its people. Sukhdev Singh Sukha and Harjinder Singh Jinda, who were later sentenced to death for assassinating the former chief of the Indian Army, General Vaidya, wrote to the President of India from the prison cell more sharply:

> You also retained the initiative and powers for Punjab's economic development. The path of development that you adopted was one-dimensional and directionless. It resulted into the imbalance of economy.
>
> Your design is to keep our industrial development at your will and never let us be self-reliant. You want to see us standing as beggars at your door. There is hardly any Agro-Industry in the Punjab. Heavy industry is totally non-existent. We want to keep our capital safe for our development, but you are exploiting us as if we were your colony. (Sukha and Jinda, 1990)

Jarnail Singh Bhindranwale, considered the fountainhead of militancy, was also not merely bothered about religion and religious demands. He was concerned about the social, political and economic problems of the Sikhs as well, but articulated these in religious terms. Most often these demands of the Sikhs and the Punjab were clubbed under one term, 'injustices'. Juergensmeyer substantiates: 'Since the larger struggle is the more important matter, these specific difficulties are of no great concern to Bhindranwale; they change from time to time. And it is no use to win on one or two points and fail on others' (Juergensmeyer, 1988: 71). Elsewhere, Pettigrew also notes that Bhindranwale's words were a religious expression of a broad-based rural discontent and anger (Pettigrew, 1984: 113). In an interview to a monthly journal Bhindranwale remarked rather simplistically that 'the Punjab and the peasant are synonyms. The former will flourish only if the latter flourishes. And only then the business of Hindu brothers will grow, otherwise it will collapse' (*Qaumi Rajniti*, 1983: 49).

It becomes clear from the above analysis that for the militants economic issues were important, be it the case of exploitation of the peasant by the big or small capitalist directly or indirectly, or by the state. The problem of poverty of lower castes and classes has also been raised by them. One issue that seems to be problematic is that they articulated these concerns in religious terminology and that too by focusing on their own community, that is Sikhs. For them Punjab is an agricultural society and agriculture is the creed of the rural people, the peasants who are no one but Sikhs. Hence the problems of rural society are the problems of the peasants and their exploitation is nothing else but an exploitation of the Punjab and the Punjabis. It is this synonymity of the region and religion with the dominant community there that seems to be at the root of the problem. It is unfortunate that in the post-Partition period the three communities—Hindus, Sikhs and Muslims—living there came to identify themselves with three different languages, namely, Hindi, Punjabi and Urdu, respectively. This kind of polarization has led to the much maligned problem of communalism in Punjab, to the detriment of one and all.

Moreover, these socio-economic demands of the Sikh militants were no imagined fears of the community. It becomes amply clear when one looks at the findings of the economists and other experts that Punjab's economy was really in a crisis. According to them, the small and marginal landholdings are becoming non-viable (S.S. Gill, 1994). The average State Domestic Product (SDP) rate is declining compared to other states like Gujarat and Haryana (Pritam Gill, 1994). The effects of the Green Revolution have been differential (Bhalla and Chadha, 1982). A serious ecological crisis is looming large due to rice-wheat cropping, hence an urgent need for crop diversification (Johl, 1986). The water table is falling at an appalling rate and the forest cover is declining.[6] Peasant indebtedness (Shergill, 1998) is rising year after year, lately becoming a major cause of their suicides.[7]

The roots of the Punjab crisis are located by Shiva in the failure of the Green Revolution: '. . . the most "successful" experiments in economic growth and development have become in less than two decades, crucibles of violence and civil war' (Shiva, 1992: 190). And Pettigrew endorses:

> The story of the rise and fall of the guerilla movement is essentially and materially a story of what happened to a community of farmers as they experienced the effects of a process of economic change known as the Green Revolution. (Pettigrew, 1995: 55)

NOTES

1. The *Anandpur Sahib Resolution* (ASR) is 'The Draft of the New Policy Programme of the Shiromani Akali Dal' adopted by its Working Committee on 16–17 October 1973 and approved by the General House on 28 August 1977 in the form of twelve resolutions. For reasons of space, it is not possible to list here all of these but it is worth mentioning that Resolution no. 1 clearly speaks for itself against the much maligned propaganda over all these years, about the separatist and disintegrationist character of the ASR. It reads:

   As such, the Shiromani Akali Dal emphatically urges upon the Janta Government to take cognizance of the different linguistic and cultural sections, religious minorities as also the voice of millions of people and *recast the constitutional structure of the country on real and meaningful federal principles to obviate the possibility of any danger to National unity and the integrity of the Country* and further, to enable the states to play a useful role for the progress and prosperity of the Indian people in their respective areas by the meaningful exercise of their powers. (p. 7) (emphasis added)

   Another feature worth noticing is its thrust on economic issues rather than religion. Without giving details it would suffice to mention that the 'Economic Policy Resolution No. 3' (pp. 8–11) occupies the most space compared to all the remaining resolutions (pp. 12–15). Very briefly it aims to put into practice Guru Nanak's idea of the Sikh way of life: 'He alone realizes the True Path who labours honestly and shares the fruits of that labour.'

   The *Resolution* declares to break the monopolistic hold of the capitalists on the Indian economy. It calls upon the central and the state governments to eradicate unemployment. It demands the redrafting of the taxation structure in favour of the poor, an international airport at Amritsar and a stock exchange at Ludhiana to accelerate industrialization and economic growth. It looks forward to parity between the prices of agricultural produce and industrial raw material. To save the farmer from exploitation the government is asked to purchase the cash crops. The SAD feels that 'the most pressing national problem is the need to ameliorate the lot of millions of exploited . . . scheduled classes'. Special funds should be earmarked for them. Finally, the SAD calls for the rapid diversification of farming, improving the land reform laws, enhancing industrialization, improving credit facilities, providing unemployment allowance and reducing the prices of tractors and tube wells.

2. There were numerous militant groups and their factions that emerged during the heydays of militancy that were subsequently clubbed under four apex bodies called the Panthic Committees. The Khalistan Commando Force was the first militant outfit, organized in 1984. This was the largest and the oldest one, constituting an independent Panthic Committee (Wassan Singh Zaffarwal). During the late 1980s, the most powerful Panthic Committee was led by Dr. Sohan Singh. It had Khalistan Commando Force (Paramjit Singh Panjwar), Bhindranwale Tiger Force of Khalistan (Sangha), Khalistan Liberation Force (Gurjant Singh Bhudhsinghwala), Babbar Khalsa International (Sukhdev Singh Babbar) and the Sikh Students Federation (Daljit Singh Bittu). That is how it got its popular name Panthic Committee Panj Membri or the Five Member Panthic Committee. The third Panthic Committee headed by Gurbachan Singh Manochahal had Bhindranwale Tiger Force of Khalistan (Manochahal), Khalistan Commando Force (Gurjant Singh Rajasthani) and the All India Sikh Students Federation (Manjit Singh). The fourth Panthic Committee of Gurdev Singh Osmanwala was of little significance.
3. The middle peasantry too got attracted towards militancy during the last years of this phase. Though clear-cut data on class lines is not available, yet as Pettigrew notes Sital Singh Mattewal owned 22 acres of land and the family of Kanwarjit Singh Sultanwind had more than 80 acres [Joyce Pettigrew (1995), *The Sikhs of Punjab: Unheard Voices of State and the Guerilla Violence,* London: Zed Books, p. 59], while Gurjant Singh Budhsinghwala had 16 acres in his name (ibid.: 120). Lately, Puri et al., have found 81.73 per cent Jat (Jutt) in a sample of 323 terrorists from the police districts of Amritsar, Majitha, Tarn Taran and Batala. Of the total sample, 64.41 per cent had 0–5 acres, 27.54 per cent had 5–15 acres and 8.05 per cent held 15–50 acres of land [Harish K. Puri, Paramjit Singh Judge and J.S. Sekhon (1999), *Terrorism in Punjab: Understanding Grassroots Reality,* Delhi: Har-Anand, pp. 60 and 62, Tables V and VI respectively]. Gurmit Singh in his study of Sur Singhwala (Amritsar district) finds that 85.86 per cent of the 41 slain militants of this village alone were Jutts and 9.75 per cent were Mazhabis. Of these, 43.90 per cent belonged to the middle class and 53.65 per cent to the low class [Gurmit Singh (1998), *Adhunikikaran ate Sikh: Ik Samajvigyanak Adhyan* (Punjabi), Patiala: Punjabi University, unpublished Ph.D. thesis, pp. 141 and 142, Tables 4.8 and 4.9 respectively]. A senior police officer informed in an interview that 60 per cent militants were Jutt Sikhs while 40 per cent belonged to the lower castes, many of whom even became the chiefs of their respective outfits.

4. The statements of the Sikh militants have been translated from Punjabi by the author. Important terms are given in parenthesis. It was taken care during translation to remain closest to the original word, in both letter and spirit, even if it meant writing bad English.
5. Bhai Ghanaiya used to provide drinking water to the injured soldiers irrespective of their religion or camp during the Sikh wars with the Muslim rulers. The Sikhs complained to Guru Gobind Singh about his nursing the soldiers of the enemy's camp. The Guru summoned him to explain his conduct. He pleaded not guilty and replied that he acted only according to the teachings of the Guru. The Guru was pleased and asked the complainants to emulate him.
6. The declining water table in the state is neither a heresy nor a projected fear of the agricultural or water scientists. More than 70 per cent blocks of the state are declared dark zones especially in the south-western region where the water table is falling at an appalling rate of more than three quarters of 1 metre per year. One can infer it from the very fact that in 1970–1 there were only 1.92 lakh tube wells but its number shot up to 11.68 lakh in 2004–5 (*Statistical Abstract of Punjab 2005*). Official figures are not available since 2005, but there has been a significant rise in their number over these years. These figures pertain both to diesel and electric operated pumping motors.
7. *Human Development Report 2004, Punjab*, p. 42. Also see K.G. Iyer and M.S. Manick (2000), *Indebtedness, Impoverishment and Suicides in Rural Punjab* (Delhi: Indian Publishers & Distributors). Vandana Shiva and Kunwar Jalees, *Farmers Suicides in India* (Delhi: Research Foundation for Science, Technology and Ecology), *Punjab De Pendu Khetar Wich Atam-Hatianwan: Ik Report* (Punjabi) (Patiala: Association for Democratic Rights, October 2000). MASR, a Chandigarh-based NGO circulated a pamphlet *Debt Trap: Farmers and Labourers* (Chandigarh: Movement Against State Repression), asking them not to 'kill themselves' and not 'sell your land . . .' in distress, rather 'go to the courts'. It advises them to understand the 'legal position on interest', its permissible limits and modes of payment, and penalty on delay, etc.

REFERENCES

Anandpur Sahib Resolution (ASR) (1973), 'The Draft of the New Policy Programme of the Shiromani Akali Dal' adopted by its Working Committee on 16–17 October.

Association for Democratic Rights (2000), *Punjab De Pendu Khetar Wich Atam-Hatianwan: Ik Report* (Punjabi), Patiala: AFDR, October.

Babbar Khalsa International (1991), 'BKI Wallon Sandesh' (Punjabi), *Jantak Paigam*, October.

Babbar, Sukhdev Singh (n.d.), *Vangaar*, Babbar Khalsa International.

Bhalla, G.S. and G.K. Chadha (1982), 'Green Revolution and the Small Peasant: A Study of Income Distribution in Punjab Agriculture', vol. II, *Economic and Political Weekly*, vol. XVII, no. 21, 22 May.

Gill, Pritam (1994), 'Punjab's Economic Development and the Current Crisis', Gopal Singh (ed.).

Gill, S.S. (1994), 'Socio-economic Contradictions Underlying Punjab Crisis', Gopal Singh (ed.).

Grewal, J.S. (1996), *The Akalis: A Short History*, Chandigarh: Punjab Studies Publications.

Gurmata (1987), Amritsar: Sri Akal Takht, 26 January.

Iyer, K.G. and M.S. Manick (2000), *Indebtedness, Impoverishment and Suicides in Rural Punjab*, Delhi: Indian Publishers & Distributors.

Johl, S.S. et al. (1986), *Diversification of Agriculture in Punjab*, Report of the Expert Committee submitted to the Government of Punjab, May.

Juergensmeyer, Mark (1988), 'The Logic of Religious Violence: The Case of the Punjab', *Contributions to Indian Sociology* (n.s.), 22 (1).

MASR, *Debt Trap: Farmers and Labourers*, Chandigarh: Movement Against State Repression.

Panjwar, Paramjit Singh et al. (1991), 'Kharku Jathebandian Wich "Bhra Maru Jang"' (Punjabi), *Punjabi Tribune*, 4 July.

——— (1991), *Ehu Benatt*, np: np.

Panthic Committee (Punj Membari) (PC [PM]), (1986), 'Khalistan Ailaan Di Dastavez' (Punjabi), *Sant Sipahi*, June.

Pettigrew, Joyce (1995), *The Sikhs of Punjab: Unheard Voices of State and the Guerilla Violence*, London: Zed Books.

——— (1984), 'Take Not Arms Against the Sovereign', *South Asia Research*, 4, 2.

Punjab Human Development Report 2004, Chandigarh: Punjab Government.

Puri, Harish K. et al. (1999), *Terrorism in Punjab: Understanding Grassroots Reality*, Delhi: Har-Anand.

Qaumi Rajniti (1983), 'Sant Bhindranwale Nal Interview' (Punjabi), Editor, August.

Ray, Niharranjan (1975), *The Sikh Gurus and the Sikh Society: A Study in Social Analysis*, Delhi: Munshiram Manoharlal.

Shergill, H.S. (1998), *Rural Credit and Indebtedness in Punjab*, Chandigarh: Institute for Development and Communication.

Shiva, Vandana (1992), *The Violence of the Green Revolution: Third World Agriculture, Ecology and Politics,* Mapusa (Goa): Other India Press.

Shiva, Vandana and Kunwar Jalees, *Farmers Suicides in India,* Delhi: Research Foundation for Science, Technology and Ecology.

Sikh Students Federation (SSF) (1989), 'Mojuda Sanwidhanak Dhanche Andar Sikhan Di Mukti Nahin Ho Sakdi' (Punjabi), *Jantak Paigam* (57), October.

Singh, Birinder Pal (2002), *Violence as Political Discourse: Sikh Militancy Confronts the Indian State,* Shimla: Indian Institute of Advanced Study.

——— (1997), 'The Logic of Sikh Militant Violence', in J.S. Grewal and Indu Banga (eds.), *Punjab in Prosperity and Violence,* Delhi: K.K. Publishers.

Singh, Gopal (ed.) (1994), *Punjab: Past, Present and Future,* Delhi: Ajanta.

Singh, Gurmit (1998), 'Adhunikikaran ate Sikh: Ik Samajvigyanak Adhyan' (Punjabi), Patiala: Punjabi University (unpublished Ph.D. thesis).

Singh Khalsa (1985), *Khalsa Raj,* np: np.

Singh, Labh and Gurjant Singh Budhsinghwala (1989), 'Eh Jang Sadi Jitt Nal Hi Mukkegi' (Punjabi), *Jantak Paigam* (58), November.

Singh, Mohinder (1978), *The Akali Movement,* Delhi: Macmillan.

Sukha, H.S. and H.S. Jinda (1990), 'Text of Sukha-Jinda's Letter to the President', *The Tribune,* 28 July.

CHAPTER 12

Burden of Tradition and Vision of Equality: Political Sociology of Jutt–Dalit Conflicts in Punjab

Ronki Ram

Caste has never been so assertive in Indian politics as it is today. Over the last few decades, however, it has entered the corridors of electoral politics with full force. Scholars, of late, have started recognizing the fact that once caste structures get politicized they help in the deepening of democracy, which in turn empowers the marginals (Yadav, 1999; Palshikar, 2004). Delivering a lecture on 'Democracy and its Critics' organized by the United Nations Foundation, Nobel Laureate Amartya Sen said: 'There is a need for caution, however, for those who believe that invocation of caste in any form in democracy is an evil force. As long as caste is invoked in speaking for a lower caste or uniting it, it is good' (*Hindu*, 16 December 2005). Such a pragmatic view of caste eclipses the common conjecture predicated on the idea that the onset of the modernity project would inevitably render invalid the institution of caste as a power index in the long run. This study is a modest attempt to understand the institution of caste in Punjab and its implications for the recent spate of Jutt–Dalit conflicts in the state.

The recent Jutt–Dalit conflicts in Punjab have exploded the myth of the casteless Sikh society. They have brought forth the latent contradictions between the landless and socially secluded Dalits, and the landowning and dominant peasant caste of Jutts. According to the *Census of India* 2001, Dalits in Punjab constitute the largest proportion of Scheduled Castes (SC) population in the country, that is, 29 per cent. Interestingly enough, Punjab has also been the only state in India where the share of Dalits owning agricultural land is the lowest (2.34 per cent). In other words, despite the fact of their being in highest proportion in the population of an agrarian state of the country, a very small number of them are cultivators. Their share in trade, industry, financial sector, health, and religious establishments in the state is also negligible (Sharma, 2003).

However, over the years Dalits of Punjab have strengthened their economic position through hard work and enterprise. Although constitutional affirmative action has played an important role in the upliftment of Dalits, in general, their monopoly in the leather business in the famous Boota Mandi in the Doaba region, and their ventures abroad have turned out to be of crucial importance in overcoming their economic hardships. In addition, they have also been politicized to a large extent by the socio-political activities of the famous Ad Dharm movement.[1] Ravidass deras[2] (religious centres) too have inculcated a feeling of self-respect among them.[3]

Thus armed with the weapon of improved economic conditions and social consciousness, the Dalits have mustered enough strength to ask for a concomitant rise in their social status. However, the Jutts have interpreted this Dalit assertion as a challenge to their long established supremacy in the state, thus sharpening contradictions between them. Dalits, who for centuries have been subjected to humiliation and untold miseries, have now learnt to say a firm no not only to the instances of violation of their human rights, but have also got ready to take up cudgels with their tormentors. Consequently, this has led to a series of violent caste conflicts over the last few years between the Dalits and the dominant Jutts. These conflicts signify the emerging Dalit assertion and its serious implications for the asymmetrically structured agrarian society of Punjab. Such conflicts are in no way a manifestation of communalism in the state. These, in fact, are signs of emerging Dalit assertion that have the potential of snowballing into serious violent conflicts, if ignored for a long time.

This paper is divided into four sections. The first deals with the regional specificities of Punjab and its impact on the phenomenon of caste discrimination. It also underlines the phenomenon of Jutt–

Dalit conflict formation in the state. The second section delves into the history of Jutt community and its links with the emergence of caste system within Sikhism. What are the patterns of caste discrimination in the Sikh society and how it forced Dalits to seek a separate identity is discussed in the third section. The fourth section documents some cases of Jutt–Dalit conflicts in pre- and post-Partition Punjab.

REGIONAL SPECIFICITIES AND CASTE HIERARCHIES IN PUNJAB

Though caste is prevalent throughout the country, it has never been monolithic and unilinear in its practice. Every region has its unique characteristics that influences its socio-political and economic structures. Thus, for a correct understanding of the phenomenon of caste and untouchability, specificities of a region hold critical importance. In the following section an attempt is made to explore the regional specificities of Punjab and their impact upon caste.

The phenomenon of untouchability was never considered so strong in Punjab as in many other parts of the country (Ibbetson, 1883, rpt. 1970). Punjab has generally been known as a 'notable exception' to the widely prevalent view of caste and untouchability in India owing to various historical factors (Puri, 2004a). But it does not mean that untouchability is alien to this part of the country. Dalits were never spared of social oppression and economic deprivations here. There are repeated references to and loud condemnations of caste-based discriminations in the teachings of Sufi saints and Sikh gurus. The social reform movements led by the Arya Samaj, Singh Sabha and Chief Khalsa Dewan further vindicate the presence of the institution of caste in the social set-up of Punjab. However, the roots of caste hierarchy were so well entrenched in the Punjabi society that reformatory measures undertaken by all social reform movements failed to weed them out.[4] However, what distinguishes Punjab from other parts of India is the material factor of caste-based discriminations as against the overall dominating pattern of the purity-pollution syndrome.

Another feature that distinguishes Punjab from other states is the phenomenon of widespread landlessness among Dalits and the absolute monopoly of Jutts on agricultural land. The hold of Jutts on land was also reinforced by the Punjab Land Alienation Act (1901) that deprived the Dalits along with other non-agricultural castes the right to purchase land. Since Punjab is primarily an agricultural state, the ownership of land assumes significance in determining social status. Nowhere in India are Dalits so extensively deprived of agricultural land as in Punjab. Despite their highest proportion in Punjab compared to other states in India, less than 5 per cent of them all cultivators (lowest in India, *Census of India*, 1991). They share only 4.82 per cent of the number of operational holdings and 2.34 per cent of the total area under cultivation (ibid.). Consequently, till recently landlessness rendered a large majority of them (60 per cent, ibid.) into agricultural labourers and made them subservient to the landowners, who invariably were Sikh Jutts. However, a significant change has taken place over the last few decades. Dalits have entered into a number of occupations traditionally considered as the mainstay of the artisan castes (Ram, 2004a). This has led to a sharp decline in the share of Dalits in the agricultural workforce in the state, down to 16 per cent in 2001 from 24 per cent in 1991 (Singh, 2005).

The hold of Jutts on the land is so strong that the lower castes are even denied access to the village common land (*shamlat*). In fact, the Dalits were never considered a part of the village, as their houses were located outside the main village. So much so that the land on which Dalit houses were built was also considered to be belonging to the Jutts (Virdi, 2003). This kept the Dalits under fear lest the Jutt landowners ordered them to vacate the land. An abysmally low share of Dalits in landownership seems to be the major cause of their hardships and social exclusion. It is also an indication of the denial of rights to them historically (Thorat, 2006). The slightest sign of protest by them for the betterment of their living conditions often provoked the Jutts to socially boycott them.[5]

The patterns of domination by the Jutts and the subordination of Dalits also distinguishes Punjab from the rest of the country in a significant way. In Punjab the scale of social measurement is starkly different. It is not based on the purity/pollution principle of Brahminical orthodoxy. Instead, it is based on the ownership of land, martial strength,[6] and allegiance to Sikhism, a comparatively new religion that openly challenges the rituals and dogmatic traditions of Hinduism and Islam. Unlike the system of caste hierarchy elsewhere in the country, the top-down rank grading of Brahmin (priest), Kshatriya (soldier), Vaishya (trader) and Shudra (menial worker) carries no meaning in Punjab where a Brahmin is not placed at the top of caste hierarchy. The Sikh Jutts, who otherwise have been Shudra as per the *varna* system, consider themselves socially superior to the Brahmins (Ibbetson, 1883, rpt. 1970; Saberwal, 1976; Tandon, 1961). In fact, in contemporary Punjab, Jutts have replaced Brahmins in terms of social domination. The ideological undercurrents of social domination based on the principles of purity/pollution and wisdom failed to hold ground in Punjab due to various historical reasons (Ibbetson, 1883, rpt. 1970; Puri, 2004a). Interestingly, the phenomenon of domination in Punjab clubbed together different sources of power (social, economic, political, religious, and numerical). These sources, in turn, were invariably concentrated in the community of Jutts. In other words, multiple identities coalesced in the Jutts to make them a dominant community. They are Jutts by caste, Sikhs by religion, and landowners by their hold on cultivation. All these different identities reinforce each other, strengthening the position of Jutts in Punjab.

Yet another factor that has further strengthened the domination of Jutts is their numerical preponderance in the Sikh religion. Their large-scale adoption of Sikhism not only rescued them from the labyrinth of their lower status in Hindu society, it also made them a powerful community in Sikh religion and politics. According to *Cencus of India*, 1881, 66 per cent of those who returned as Sikhs were Jutts. The second largest community within Sikhism was that of Tarkhans/Ramgarhias (carpenters) who constituted 6.5 per cent of the total Sikhs in Punjab. Next to them were Chamars/Ramdasias (5.6 per cent), followed by Chuhras/Mazhabis (2.6 per cent). If clubbed together these two outcaste groups (Ramdasias and

Mazhabis) become the second largest group (8.2 per cent) of Sikhs in Sikh society. Thus, the numerical prepondrance of Jutts in religion combined with their martial and self-willed nature and monopoly on land 'elevated them well above their humble origins'.

Such a combination and reinforcement of multiple identities and their concentration in the community of Jutts is, however, conspicuous by its absence among the Dalits, which weakens their collective strength and unity. Dalits in Punjab are scattered in multi-identities. Under the impact of Sikhism, while Jutts have enhanced their social status and achieved spiritual coherence, the same could not happen in the case of Dalits who have remained divided in different religious orders. Dalits are found in all major religions in Punjab. Their presence in Hinduism, Sikhism, Islam and Christianity not only proves the presence of the institution of caste in all these religions, but also weakens the chances of solidarity among them.

The subjugation of Dalits got further deepened during the course of the Green Revolution in Punjab that transformed the traditional subsistence character of agriculture into commercial and mechanical farming. The market-oriented agriculture pattern in the post-1960s favoured the landowners, which further marginalized the Dalits and widened the existing divisions between them and the dominant peasants. Interestingly, it was also during this phase that a new middle class of educated Dalits emerged that coupled with the rise of the Ambedkarite movement in the region lead to the rise of Dalit consciousness.

The emergence of Dalit consciousness induced the Dalit agricultural labourers to ask for higher wages in the rural areas, especially in Doaba. Their struggle for higher wages often employed pressure tactics of refusal to work unless the landowners increased their wages. In fact, it was during this very phase of transition in the agrarian economy of Punjab that the process of Dalit immigration to Europe, North America, and the Gulf got streamlined. This process, however, coincided with the influx of migrant labour from Bihar and eastern Uttar Pradesh into Punjab that further sharpened contradictions between the dominant peasants and the landless Dalits, as the migrants provided labour cheaper to the locals. Moreover, the changed cropping system under the Green Revolution squeezed the extent of farm labour to a few peak periods—paddy transplantation, paddy harvesting-cum-threshing, and wheat harvesting. The traditional agriculture providing labour round the year was changed into a commercial enterprise that did not offer more than 75 days of work annually (Singh, 2001). It forced the Dalits to seek employment in other sectors.

Thus, Dalit labourers sandwiched between the influxes of cheap migrant labour and mechanized farming began to look for jobs in sectors other than agriculture. The alternative job opportunities reduced the dependence of Dalits on landowners. The social mobility of the new middle-class Dalits coupled with their relative emancipation from economic dependence on the landowners led to the emergence of Dalit assertion in Punjab. The sustainability of this assertion drew strength from the politicization of caste on the one hand and from the failure of asymmetrical caste structures to accommodate Dalits into its social space as equal citizen, on the other (Judge, 2006). This new form of Dalit assertion and its recent exhibition in the form of Jutt–Dalit clashes in the villages demands a serious social science enquiry.

Yet another feature that distinguishes Dalits of Punjab from their counterparts in other regions of the country is their community-wise heavy concentration in some pockets of the state. Dalits in Punjab have been categorized into 38 castes out of which more than 80 per cent of the total SCs population belongs to two major castes of Chamars (leather workers) and Chuhras (sweepers). These two castes consist of four sub-castes—Mazhabi (30.7 per cent), Chamar (25.8 per cent), Ad Dharmi (15.9 per cent), and Balmiki (11.1 per cent). The Chamars include Ad Dharmi, Juttia Chamar, Rehgar, Raigar, Ramdasias and Ravidasias. The Chuhra caste clubs together Balmiki, Bhangi and Mazhabi. The Chamar caste is largely confined to Doaba (comprising Jalandhar, Hoshiarpur, Kapurthala, and Nawanshahr districts between the two rivers Beas and Sutlej). The Chuhra caste is mainly concentrated in the smaller Majha and the larger Malwa regions. At the district level, Mazhabis are largely concentrated in Ferozepur, Gurdaspur, Amritsar, Faridkot, Mansa and Bathinda districts. Apart from their heavy concentration in Doaba, Chamars are also numerous in Gurdaspur, Rupnagar, Ludhiana, Patiala and Sangrur districts. Among the Chamars, Ad

Dharmis far outnumber other SCs in Jalandhar and Hoshiarpur districts in rural as well as urban areas. Mazhabis in Chuhra caste outnumber other SCs in Faridkot and Ferozepur districts (for details Gosal, 2004). Though traditionally they have been condemned as polluted and impure because of their occupational contact with animal carcass and hides, Chamars have *chandravanshi* lineage and are considered highest among the SCs in Punjab (Deep, 2001).

Ad Dharmi and other Chamar sub-castes are not only numerically preponderent in Doaba, they also happen to be most resourceful in comparison to all other castes among the SCs of Punjab. Chamars and Ad Dharmis of this region are ahead of all other Dalit castes in almost all spheres. 'Ad Dharmi Chamars are on the top of virtually every parameter—education, urbanization, jobs, occupational change, cultural advancement, political mobilization, etc.' (Puri, 2004: 4). The famous Ad Dharmi movement of the 1920s also emerged in Doaba. In the early 1930s, some Ad Dharmi Chamars established a prosperous leather-business township (Boota Mandi) on the outskirts of Jalandhar City. They were also among the early supporters of the Ad Dharm movement. Seth Kishen Dass, a leather tycoon of Boota Mandi, who won the 1937 assembly election from Jalandhar constituency in Punjab, financed the building housing the headquarters of the Ad Dharm Mandal in Jalandhar city. Nowadays, this building houses Guru Ravidass High School and a sewing centre. It is again from this caste of Doaba that maximum emigration took place to Europe, North America and the Middle East. The Ad Dharmis abroad have not only excelled in business and skilled labour jobs, they have also established a strong networking of social organizations—Ravidass Sabhas, Ravidass Gurdwaras—and hold international Dalit conferences throughout Europe and North America.

SIKHS, JUTTS AND CASTE

Punjab is a Sikh majority state, the Sikhs constitute 63 per cent of its total population. About 72 per cent of the Sikhs live in villages where caste as an occupational division of labour constitutes an integral part of routine social life (Kaur, 1986). Although the Sikh doctrine negates the institution of caste, the same is not true in social practice (Puri, 2003). In the Punjab censuses from 1881 to 1931, more than 25 castes were recorded amongst Sikhs, viz., Jutts, Khatris, Aroras, Ramgarhias, Ahluwalias, Bhapas, Bhattras, Sainis, Lobanas, Kambohs, Ramdasias, Ravidasias, Rahtias, Mazhabis, and Rangretas (Verma, 2002: 33). Out of these, eleven castes—two agrarian castes (Jutt and Kamboh); two mercantile castes (Khatri and Arora); four artisan castes (Tarkhan, Lohar, Nai, and Chhimba); two outcastes (Chamar and Chuhra); and one distiller caste (Kalal) —remain the principal constituents of the Panth (McLeod, 1996).

The outcaste Sikhs, popularly known as Dalit Sikhs, are divided into two segments: Mazhabis and Ramdasias.[7] The Dalits whose occupation is scavenging and cleaning are called Mazhabis. 'Mazbi means nothing more than a member of the scavenger class converted to Sikhism' (Ibbetson, 1883, rpt. 1970: 294). Some sweepers who embraced Sikh religion are also called Rangretas. However, in spite of Mazhabis and Rangreta Sikhs' meticulous observance of the Sikh religious principles, they are not considered equal by the upper caste Sikhs who refuse to associate with them even in religious ceremonies (ibid.). In other words, even after converting to Sikhism, they have not been relieved of the taint of hereditary pollution. The other segment of Dalit Sikhs consisting of Ramdasias is also known as Khalsa Biradar. They are Chamars who have converted to Sikhism. Most of them are Julahas (weavers). They are often confused with Ravidasia Chamars who are mostly engaged in the occupation of leatherwork (ibid.: 300).

Mazhabis/Rangretas and Ramdasias are not equal to the Jutts, Khatris and Aroras in Sikh society. Their status is lower even to Ramgarhia, Ahluwalia and Bhapa Sikhs. Thus, change in the caste titles of the Dalits after their conversion does not make any difference to the dominant castes. The latter continue to identify them by their earlier titles—Chuhras and Chamars. Though Mazhabis or Rangretas have abandoned the occupation of scavenging, they still are clubbed with Chuhras (ibid.). As far as the Dalits are concerned, they too continue to observe caste among them even after their conversion to Sikhism. Ramdasia Sikhs consider themselves superior to Mazhabi and Rangreta Sikhs. Although Ramdasias and Ravidasias have originated from Chamars, the former (Sikh) consider them superior to the latter (Hindu) (ibid.).

In the Sikh caste hierarchy, Jutts claim to occupy the top position (Singh, 1977). To quote Pettigrew, an anthropologist who has done intensive fieldwork on the Sikh Jutts: 'All Jutts alike are brought up to be proud irrespective of what they possess in terms of education, wealth or power. No Jutt defines himself as subservient and none can actually be trampled upon' (Pettigrew, 1978: 20). Mostly concentrated in villages, the Jutts are primarily landowners and agriculturists and are also widely considered to be the backbone of the Punjab peasantry. 'So close has become the connection of the Jutts with peasant-agriculture in the Punjab that, besides being a caste-name, the word Jat can mean an agriculturalist and *Juttaki* similarly can mean agriculture' (Habib, 1996: 97). By virtue of their hold on the land they are popularly known as the dominant peasant caste in the state. 'The Jutt might be employed as a school teacher, or service in the military but he sees his primary role as that of an agriculturist; his connection with land is what he holds most dear and what identifies him' (Kaur, 1986: 233). The Jutt have also diversified into transport business and consider employment in the armed forces highly prestigious.

The Jutts in Punjab are also considered the backbone of the Sikh community. Although all the ten Sikh gurus belonged to the Khatri caste, traditionally the majority of their followers have come from the Jutt caste (ibid.). In the *misl* (military bands) system of the eighteenth century, leadership was largely under the control of Jutts and 'eventually it was a Jat (Jutt) *misldar*, Ranjit Singh, who secured total ascendancy'(McLeod, 1996: 18). The overwhelming majority of Jutts (since 1962) in the leadership of the Shiromani Akali Dal, the main political party of the Sikhs, has made it 'virtually a Jat (Jutt) political party' (Puri, 2004a: 10).

The Sikhs are identified by their appearance based on the five symbols, *kirpan* (steel dagger), *kara* (steel bangle), *kachchh* (short breeches), *kangha* (comb) and *kesh* (uncut hair) that they wear in accordance with the *Rahatnama* (Sikh code of conduct). 'However, Sikh Jutts are generally liberal in observance of the *Rahatnama*. Majority of them trim beard, cut hair, and many often smoke or chew tobacco. They rarely visit Gurdwaras' (Kaur, 1986: 222–3). In spite of their lackadaisical approach towards the Khalsa discipline, Sikh Jutts in their own eyes and in those of others remain Sikhs. 'For other castes it is very different. If a Khatri shaves he is regarded as a Hindu by others and soon comes to regard himself as one' (McLeod, 1996: 98). The Sikhs who strictly followed the *Rahatnama* belong to the lower class of north Punjab (Singh, 1953).

The Khalsa symbols were considered to be associated with the influx of Jutts into the Sikh religion during the eighteenth century (McLeod, 1996; Pettigrew, 1978: 25). However, with the passage of time, the symbols became permanent part of the Khalsa discipline in 1699. Since these symbols were part of the 'Jutt cultural patterns' much before the entry of Jutts into Sikhism, their adherence by the Jutts could not become an identification mark of their being Sikhs. Even before they became Sikhs they kept uncut hair, wore a thick *kara*, and a turban, as a measure of protection in warfare. Hence, the importance of these symbols did not make much difference to them after their becoming Sikh. So, in their case it was not the adherence to these symbols that made them look like Sikhs. They remained Sikhs even without wearing these very symbols. In other words, the entry of Jutts into Sikh religion did not dilute their 'caste identity'. On the contrary, it got further strengthened. Jutts considered themselves saviours of the Sikh religion: they defended it militarily throughout its entire turbulent history. In the words of Pettigrew:

> Each Jutt felt tremendous pride that it was *his* section of the community that had built up the military organization which led to the establishment of Sikh rule in the Punjab. He felt that prestige lay with the Jutts because of this. (Pettigrew, 1978: 41, emphasis in original)

The Jutts often treated other castes as timid and incapable of defending themselves. They called Aroras *kariar* (coward), and commonly applied the term *bhapa* (which carries a perceptible degree of opprobrium) to Khatris and Aroras who migrated from the Pothohar area (McLeod, 1996; Pettigrew, 1978).

The Jutts are generally considered to be of Indo-Scythian stock, and are said to have settled in the Indus Valley, especially in central Sind, in the seventh century (Habib, 1996). They were ruled over by the Brahmana dynasty of Chach that imposed harsh constraints on them (ibid.: 95). Their appearance became apparent in Punjab by the beginning of the eleventh century (Ibettson, 1883, rpt. 1970; Habib, 1996). The entry of Jutts into the province of Punjab must have been based on their migration from Sind (Habib, 1996). However, for a period of 400 years between the eleventh and sixteenth centuries there is no account of them in the chronicles of Punjab. This simply shows their insignificance in the Punjab society. Al-beruni, whose historical account covered the period of eleventh century, designated them as 'cattle-owners, low Shudra people' (quoted in ibid.). They were known as people 'of an unfeeling temper' and 'hasty disposition', who were free from the dichotomies of 'small or great' and 'rich or poor'.

References to the Jutts began to surface again after a long gap of four centuries in 'the *Ain-i-Akbari* and its record of *Zamindar* castes, compiled about 1595' (ibid.: 96). During the four centuries of their insignificances the Jutts must have expanded and metamorphosed from a pastoral to an agricultural community in Punjab (ibid.). This was probably also the period during which cultivation expanded substantially in Punjab. The introduction of Persian Wheel, reiterated Irfan Habib, was the main driving force behind the 'critical change in the agricultural situation of the Punjab' (ibid.: 98). The expansion of agriculture in Punjab might have led to massive shift among the Jutts from being a pastoral to settled agricultural community. It is safe to say that it might have also elevated their social status in the political economy of rural Punjab. It would not be out of context to say that what Green Revolution was to the post-1960s Punjab, the introduction of Persian Wheel was to the Punjab of the sixteenth and seventeenth centuries. In both cases, it was the Jutt community that remained the main beneficiary of the transformation process in the rural settings. But how did the pastoral Jutt community get transformed into a settled agricultural community and establish its control over land? This question cannot be answered simply by asserting that since the pastoral Jutts were tending cattle, and cattle are generally related with agriculture, so they adopted this occupation. Agriculture is not merely an occupation, it is also an asset that bestows on the owners of land the special status of a *zamindar*. Jutts' hold on agricultural land probably made them an important community. In the sixteenth century when many of the Jutts turned to cultivation, they '... were not only entirely peasants but, in so many localities of the Punjab, also *Zamindar*.'... (Habib, 1996: 99; Ibbetson, 1883, rpt. 1970: 103). In fact, it was their hold on the land that became a marker of their 'Jutt identity'. The Jutt and agriculture, thus, became synonymous.

However, the Jutts' improved economic conditions failed to push them up the caste scale within the Hindu social order. Thus, to escape the oppressive and suffocating structures of Hindu society the Jutts of Punjab embraced Sikhism—a newly emerged religion, free from the hierarchies of caste and gender (Habib, 1996; McLeod, 1996). They saw in this new religion a hope and a promise to win over the dilemma of incommensurability between their improved economic position and humiliating social status. Since the Jutts constituted a large segment of the population of Punjab, their entry into the Sikh religion quickly made them the preponderant community. In fact, the large-scale entry of Jutts into the Sikh religion had not only expanded the base of this new religion, it had also seriously impacted its social outlook. It introduced elements of militancy and caste in its organization. The militant outlook of the Panth (Sikh community) especially after the martyrdom of the fifth guru Arjan Dev is generally attributed to what McLeod called the preponderance of the 'Jutt cultural patterns' within Sikhism. The preponderance of such patterns also turned Jutts into a dominant caste within the very religion that purged them of taint of their lower caste status. In due course of time they came to be known as the dominant caste in the whole state. So much so that 'Punjabi culture and identity is seen in terms of Jutt culture and identity only' (Jodhka, 2006: 13). In the words of Grewal:

> Although due to the present agricultural crisis in Punjab this community is in an unfortunate and painful condition, but still if anybody asks who is most powerful in Punjab, we would have to acknowledge that these directionless, Jutt Sikh families of Punjab, that is committing suicide [*sic*], are the ruling class here. (Grewal, 2006: 16)

The transformation of Jutts from a pastoral community into an agricultural one, and their allegiance to the Sikh religion reveals an interesting case of the empowerment of a lower caste and the role of religion in that regard. In fact, what the Jutts were fighting for in the seventeenth and eighteenth centuries, the Dalits of Punjab seem to have been struggling for the same over the last few decades in contemporary Punjab. They have been fighting for an equal share in the sources of power and a respectable status in society. Though they have gained some elevation over the years in their economic status due to the constitutional affirmative action and ventures abroad, their lower social status remains intact. Unlike the lower caste Jutts of the seventeenth and eighteenth centuries, they have failed to overcome their social disability by embracing Sikhism. The Mazhabis of Punjab is a case in point.

> The Mazbis take the *pahul*, wear their hair long, and abstain from tobacco, and they apparently refuse to touch night soil, though performing all the other offices hereditary to the Chuhra caste. . . . But though good Sikhs so far as religious observance is concerned, the taint of hereditary pollution is upon them and Sikhs of other castes refuse to associate with them even in religious ceremonies. (Ibbetson, 1883, rpt. 1970: 294)

Ironically the Jutts failed them in their efforts to throw away the shackles of caste hierarchy.

However, there is one major factor that distinguishes the Dalit case from that of the Jutts in the formative years of their struggle for the improvement of their social status. Jutts were cultivators, landowners, nonchalant and a martial race. They also outnumbered other communities by their numerical strength within the Sikh religion. Moreover, the contradiction between the principal communities of the Khatris—the community to which all the ten gurus belonged and also the one which provided the initial following to the Sikh religion—and the Jutts was never sharp. Whereas the Jutts remain a rural community heavily committed to agriculture, the Khatris are essentially urban-based and a mercantile community (McLeod, 1996). To quote McLeod: 'Unlike the Jutts the Khatris have never shown any interest in Sikh identity as a means of enhancing social or ritual status . . .' (ibid.: 99). Nor they ever opposed such attempts of the Jutts. Thus, the Jutts of Punjab never faced any opposition from the Khatris in their efforts towards social upliftment.

But in the case of Dalits, the situation is entirely different. Dalits in Punjab are posited in direct confrontation with Jutts over the struggle for social justice and dignity. Unlike the Jutts of the eighteenth century whose opponents (Khatris) were in no way directly entangled with them in their occupation (agriculture), some Dalits of Punjab are still tied with Jutts in the sector of agriculture. It is in this context that Jutts, the landholders and Dalits, the landless agricultural workers, find themselves in direct confrontation. But there are many Dalits in the state who have improved their economic conditions by dissociating from their caste occupations and distancing themselves from agriculture. Some of them have joined government services, gone abroad and established their own small-scale servicing units (carpentry, barber shops, blacksmithy, etc.) (Ram, 2004a). They have not only improved their economic status, but have also liberated themselves from the subordination of Jutt landowners. Now, they no longer feel obliged to respect their erstwhile masters (Jutts). Thus changed economic relations have not only improved their economic status, but also propelled them to aspire for a commensurate social status. This is what has pitted them against the Jutts, who find it hard to digest any attempt, which would dilute their dominant position in the rural society. The Dalits's struggle for equal social status, thus, has led to violent caste conflicts between them and the Jutts and has all the probability of escalating into many more clashes in the near future.

JUTTS AND CASTE DISCRIMINATION

Caste discrimination in Punjab is unique in comparison to its observance in other parts of the country. The Brahminical tradition of social stratification has never been so effective here. The word Brahmin did not carry a sacerdotal connotation in Punjab. It was used rather derogatorily. The down play of Brahmins by the Sikh Jutts might have diminished the purity-pollution practice to the benefit of the Dalits (Saberwal, 1973). However, it did not in any way help the Dalits to improve their social status.

The centre of power in Punjab revolves around the axle of land. Much of the land is owned by the Sikh Jutts. Although the SC constitute a high proportion of the population (29 per cent) in comparison to the all India average of 16.3 per cent, their share in ownership of land is negligible. Their landlessness has forced them to depend on the landowning caste in the absence of alternative jobs in the agrarian economy of the pre-Green Revolution Punjab. Since cultivation required the services of Dalits in its various operations, it was not feasible to strictly follow the system of untouchability based on the principle of purity-pollution. It does not mean that the Dalits were not discriminated in Punjab. They were discriminated. However, the context of their discrimination was different compared to other parts of India. The practice of untouchability in Punjab was based on the scheme of keeping the Dalits bereft of landownership and political power. They were forced to confine to their lowest status in rural Punjab lest they dare ask for a share in the power structure (Puri, 2003). In other words, despite the absence of purity-pollution syndrome, the presence of deep asymmetrical structure of power in the agrarian village economy has subordinated the Dalits to the landowning upper castes (Jodhka, 2002).

The villages are divided into upper caste localities, and the Dalit settlements are located invariably on the side towards which the dirt of the village flowed. The Dalits are not allowed to build pukka (concrete) houses because the land on which they live, does not belong to them. In the villages they are often involved in unclean occupations—carrying and skinning dead animals, scavenging and working as attached labourer—as *siri.* Nowadays, such work is performed on non-*jajmani* basis. In Malwa region, there are many Dalits who are still working as *siris.* According to a latest study of 26 villages in Malwa, 21 had Dalits working as *siris* (Jodhka, 2002). Another study found six Jutts working as *siris* with other Jutts in a village in Sangrur district (Singh, 2001). However, the situation is entirely different in Doaba where the majority of Dalits have dissociated themselves from such types of menial works. Although the

Dalits interacted with the Jutt Sikhs, being agricultural labourers and *siris*, they used to keep their own tumblers and plates to take meals or tea or water from the upper caste Sikhs.

> The upper caste Sikhs are a separate identity and like the upper caste Hindus they also follow the ideology of a graded human society . . . the Sikhs may take food with the Dalit-Sikhs in Gurdwaras, but they have no bond of fraternity with them. (Singh, 2002: 333)

To quote Singh again: 'the impact of Hinduism and caste is visible on the adherence of Guru Nanak and they monopolized Sikhism and could not accord an equal social status to the lower caste Sikhs in Punjab' (ibid.). The Dalit Sikhs in Punjab are cremated in separate cremation grounds along with their counterparts in the Hindu religion. Even in some villages the land meant for cremation grounds in the *shamlat* have been grabbed by the upper castes. In such a recent case the dominant caste persons of Todder Majra (Mohali) grabbed the cremation grounds of the Dalits (*Desh Sevak*, 2 January 2005). This shows that the social position of the Dalit Sikhs in Punjab is no better than that of other Dalits elsewhere in the country (ibid.).

The Dalit Sikhs did not get equal treatment in the Gurdwaras of the upper caste Sikhs. 'Mazhabis were forbidden to enter the Golden Temple for worship; their offering of *karah prasad* was not accepted and the Sikhs denied them access to public well and other utilities' (Pratap Singh in Puri, 2003: 2697). The Dalit Sikhs were not allowed to go beyond the fourth step in the Golden Temple and the members of the fourfold *varnas* were instructed not to mix with them (Oberoi in ibid.). Evidence of untouchability against Dalit Sikhs is vividly reflected in a number of *Gurmatas* (resolutions) adopted by the Shiromani Gurdwara Prabhandak Committee from 1926–33 (ibid.). Although removal of untouchability figured in the Singh Sabha movement, no strenuous effort was made in that direction. 'It was not surprising. For the Jutts, who composed 70 per cent of the Akalis, and other high castes, caste equality or removal of untouchability was contrary to their disposition for social domination and hierarchy' (ibid.). This has forced the Dalit Sikhs to establish separate gurdwaras, which, in turn, has further led to the strengthening of the already existing caste divisions amongst Sikhs[8] (ibid.; Jodhka, 2002; Muktsar, 1999 and 2003). Moreover the observance of caste prejudices against the Dalit Sikhs has compelled them to 'search for alternative cultural spaces' in a large number of *deras*, sects, and *dargahs* of Muslim *pirs* and other saints (Puri, 2003: 2700).

However, for the last few decades the Dalits of Punjab have 'discovered the right remedy to cure their wounded psyche' in the famous Dera Sant Sarwan Dass situated at village Ballan (Jalandhar) (Rajshekar, 2004: 3). This *dera*, popularly known as Dera Ballan, has become a paragon of the Ravidass movement in north-west India. It has been playing a leading role in promoting cultural transformation and generating social consciousness among the Dalits of the region. The *dera* has a library in its premises, publishes a tri-lingual weekly, distributes free Dalit literature, honours Dalit scholars, runs a model school and a hospital for the service and upliftment of the downtrodden. It has made concerted efforts for the construction of a separate Dalit identity. The saints of Ballan developed their own religious symbols, flags, prayers, dress, salutations and rituals of worship. Of all the major contributions that the Dera Ballan made, the construction of a mammoth temple at Shri Guru Ravidass' birthplace—Seer Goverdhanpur in the vicinity of Varanasi—is the most significant. This temple has acquired, perhaps, the same importance for the Dalits as Mecca for Muslims and the Golden Temple for Sikhs.

SOCIAL EXCLUSION AND VIOLENCE IN COLONIAL PUNJAB

Dalits of Punjab faced stiff opposition and became victims of physical violence at the hands of the dominant castes during their struggle for dignity and equality in the colonial period. They were, said an eyewitness: '. . . chased everywhere and hounded out of bounds of towns and villages by the Hindus and quite often they had to hold their meetings and conferences in open fields. One such incident also took place at Una' (Pawar, 1993: 77). They were also denied entry into meadows and common lands to fetch fodder for their cattle, access to open fields to answer the call of nature, and were interned in their houses by the Sikhs and Hindus for no other fault than that of their being registered as Ad Dharmis in the census of 1931. In Ferozepur district, two Chamars were burnt alive because they registered themselves as Ad Dharmis (Chumber, 1986). In Layalpur district, the innocent daughter of an Ad Dharmi was murdered. In Nankana Sahib, the Akalis threw ash into the *langar*, food prepared in free community kitchen for the participants in an Ad Dharm meeting. In Dakhiyan-da-Prah (Ludhiana), some Sikh boys abducted Shudranand from the dais of the *achhuts*' (Dalits) public meeting. In Baghapurana, many *achhuts* were beaten up and their legs and arms broken (Bakshi, n.d.). In many villages of Ludhiana, Ferozepur and Lyallpur districts, *achhuts* who were living in villages where the Jutt-Sikhs or Muslims were in a dominant position were boycotted for two months. The Sikh Jutts had compelled the *achhuts* to record themselves as Sikhs. However, 'despite repression and intimidation the *achhuts* did not give in and recorded Ad Dharm as their religion' (ibid.: 54–6). At Ghundrawan (Kangra), the Rajputs even smashed the pitchers of Ad Dharmi women who were on their way to fetch water. When denied water from the village pond, Ad Dharmis had to travel 3 miles to a river for water. The ongoing torture at the hands of Rajputs ultimately compelled them to leave the village to settle in Pathankot. It was only after the interference of Sir Fazal-i-Hussain, on the request of Mangoo Ram[9] that their grievance was looked into and eventually they got rehabilitated in their native village.

In the face of opposition by the upper caste Hindus, Sikhs and Muslims, the leaders of Ad Dharm had a tough time proving to the Lothian Committee that they were neither Hindus, Sikhs, Muslims nor Christians (Piplanwala, 1986; Ahir, 1992). The Sikh representatives claimed that since many *achhuts* believed in the *Guru Granth Sahib* and solemnized their marriage ceremonies in accordance with the Sikh customs, half of their population should be declared Sikh and the other half as Hindu. Likewise the Muslim representatives told the Lothian Committee that since some *achhuts* perform *namaz* (Muslim prayer), keep *rozas* (a month-long fast) and bury their corpses in cemeteries, they should be divided equally between Hindus and Muslims. Similarly, the Hindu representatives stressed that since the *achhuts* believed in Vedas and perform their marriage ceremonies in accordance with the Hindu customs no one except they can claim their allegiance. Above all, Lala Ram Das of the Dayanand Dalit Udhar Mandal (Hoshiarpur) and Pandit Guru Dev of Achhut Mandal (Lahore) informed the franchise committee that there was no untouchable in Punjab. According to them, the untouchables were the backward 'classes' of Hindus who were made at par with the rest through the performance of *shuddhi*. Hence, no separate treatment for the untouchables in Punjab was required.

> Untouchables generally were being subjected to strong pressures by Muslims, Hindus, Sikhs and others, each community seeking to pull them into its own fold, at least for the day of the census: it was common then to seek to influence census results as a prelude to political claims. (Saberwal, 1976: 52)

Thus the Dalits were put to severe hardships and violence for carving out an identity for themselves and asserting for their rights in the colonial period.

SOCIAL EXCLUSION AND VIOLENCE IN CONTEMPORARY PUNJAB

Atrocities against the Dalits continued to afflict them even after India became independent. Unfortunately the frequency and magnitude of such atrocities increased after the 1960s in the wake of the Green Revolution in Punjab. Rarely a day passed when Dalits were spared of a social boycott by the Jutts in the villages. After the much-publicized violent conflict in Talhan (Jalandhar), Punjab has witnessed a series of similar cases. There is a similarity of pattern. In almost all the conflicts, social boycott was clamped on the Dalits who were asserting for equal rights in the village power structure. Pandori Khajoor (Hoshiarpur), Bhattiạn Bet and Aligarh near Jagraon (Ludhiana), Talhan, Meham and Athaula (Jalandhar), Patteraiwal (Abohar), Jethumajra and Chahal (Nawanshahr), Domali and Chak Saboo (Kapurthala), Abul Khurana near Malout (Muktsar), Dalel Singhwala, Kamalpur and Hasanpur (Sangrur) and Jhabbar (Mansa) are among the most prominent cases of Jutt–Dalit conflicts in the state. In the following section Talhan and Meham conflicts are briefly discussed to analyse the underlying causes of the caste-based oppression in contemporary Punjab.

TALHAN

The Talhan conflict was based on the issue of Dalit representation in the management committee of Gurdwara Shaheed Baba Nihal Singh. Dalits were denied access to the management committee of this gurdwara built on the tomb of Baba Nihal Singh, a local carpenter (backward caste) who died while laying *gandd* (wooden wheel) at the base of a well. Since Baba was popular for his expertise and died while working for a public cause, his death was not considered an ordinary event. The villagers declared him a *shaheed* (martyr). They constructed a small *smadh* (tomb) at the place of his cremation. Another *smadh* was also built nearby in the memory of Harnam Singh, an aide of the Baba who for years cared for his *smadh.* To celebrate the martydom of Baba, the villagers started organizing an annual fair at his *smadh.* The popularity of the Baba began to attract a large number of devotees who brought offerings, mostly in cash. Subsequently, the *smadhs* were converted into shrines. In due course, another structure—a gurdwara—was raised between the *smadhs* and the Sikh scripture was placed there. The whole site, including the two *smadhs,* thus, turned into a gurdwara.

The conversion of *smadhs* into a gurdwara was widely seen as an attempt to grab the large offerings made by Jutts of Talhan and the adjoining villages. The Jutts (25 per cent) who control most of the agricultural land and until recently enjoyed unquestioned domination in the social and political life of the village established their control over the gurdwara through its management committee. This committee manages approximately Rs. 50 million ($1.1 million) annually, which the gurdwara receives in offerings from Punjabi diaspora and local devotees (Philip, 2003). There may be a difference of opinion on the exact amount of offerings, but as Philip puts it: 'There is an agreement that the coffers in the gurdwara have been overflowing with cash. Small wonder that anybody who is a somebody in the village wants to be a member of the gurdwara management committee' (ibid.).

Despite being a majority (72 per cent) in the village, the Dalits were kept out of the membership of the Gurdwara Management Committee. The numerically dominant Dalits, majority being landless, have achieved a considerable degree of mobility and autonomy over the last few decades. They have diversified into non-agricultural sectors and also found employment abroad. Their numerical strength has also added to their importance in the electoral politics of the village. Consequently, they started vociferously demanding a share in the power structure at different levels of Punjabi society, which hitherto has been dominated by the landholding castes, particularly the Jutts. These demands for a share in the local power structure led to Jutt–Dalit clash in Talhan.

The Dalits here employed every available method to seek entry into the Gurdwara Management Committee. They requested the Jutts to give them due share in the membership of the Committee in proportion to their population in the village. The Jutts refused. Then, in 1999, the Dalits approached the local administration and the court of law. But the dispute remained unresolved. However, the Dalits continued their struggle to acquire membership. This ultimately led to a fight between them and the Jutts in January 2003. Subsequently, the Jutts publicly announced their social boycott. The non-Dalit residents were asked to sever social and economic ties with the Dalits. The Jutts stopped visiting the shops run by them. They also banned the entry of Dalits in their fields. They were not allowed to defecate in the fields, thus forcing them to do so by the roadside.

To fight against the social boycott and for representation in the committee, the Dalits organized a Dalit Action Committee (DAC) under the leadership of L.R. Balley, a prominent Ambedkarite of the region. The DAC organized sit-ins and hunger strikes in the village and in Jalandhar city. Repeated appeals by the DAC to the Punjab government for legal action against the boycotters failed to move the administration (Singh, 2003). On 5 June 2003, the conflict took a violent turn and soon spread to the adjoining areas. Boota

Mandi, a suburb of Jalandhar city, became the epicentre of the violence. It was here that an Ad Dharmi, Vijay Kumar Kala, fell victim to police firing, an event that propelled Talhan and Boota Mandi onto the national scene. These places were virtually converted into garrisons and Talhan was sealed off for a couple of days.

The pressure of Dalit assertion compelled the government to resolve the conflict without delay lest it become a serious political issue with wider implications. Moreover, it also cautioned the government to take necessary steps to prevent the victimization of Dalits in other parts of the state, lest they replicate Talhan. Although the district administration and the police controlled the violence, it took the contending parties 18 days to reach a compromise, and another two months for the agreement to come into effect. Eventually the Dalits of Talhan succeeded in securing representation in the Gurdwara Management Committee. Though the Talhan conflict was a case of local Dalit upsurge, it has set a historic precedent in Punjab through Dalit assertion (for details Ram, 2004b).

MEHAM

Meham presents another case of recent Jutt–Dalit conflict, and a vindication of the existence of the institution of caste in Punjab. Meham has a total population of 1,967 out of which 893 (45 per cent) are Dalits. Most of them belong to the Balmiki caste. Ad Dharmi, another Dalit caste, constitutes 20 per cent of the population (Judge, 2006). The Sikh Jutts are also about 20 per cent. The Jutts, Balmikis and Ad Dharmis each have their own gurdwara. In fact, the Jutts have two gurdwaras. Baba Khazan Singh Udasi Dera (site of dispute) is the fifth shrine in Meham. As has been the case in majority of the villages in Doaba, the Dalits in Meham have also diversified into various non-agricultural sectors. This has not only helped them abandon their customary caste-based occupations but also liberated them from their dependence on land, and hence Jutts. Despite all this they have failed to raise their social status in the eyes of the Jutts. This is the source of tension between them.

Though the context of Meham conflict is different from that of Talhan, the patterns and forms of oppression of Dalits are the same in both cases. In Talhan, the Jutts denied entry to Dalits in the management of the gurdwara. In Meham, the Jutts forcibly took over the control of the Udasi Dera being looked after by Ad Dharmis for the last six decades. The Jutts replaced all the Udasi symbols with that of the Khalsa, and also objected to offerings of liquor and its distribution as *prasad* among devotees as it violates the Sikh code.

The Ad Dharmis retorted that the tradition of offering liquor at *smadh* is no violation of the Sikh code as the Dera was never a site of Sikhism. They reiterated that the issue of Sikh code was raised in 2003 when the Sikh Jutts placed the *Guru Granth Sahib* in the premises of the Dera. Moreover, the presence of *mazaars* (graves) in the precincts of the Dera rule out the possibility of its being a gurdwara. In the Talhan conflict too, the Dalits also raised the same argument over the dispute of the grave of Shaheed Baba Nihal Singh. Jutts' control of the Dera, however, could be attributed to the rising cost of land and tremendous increase in the donations and offerings there from the Punjabi immigrants in Europe, North America and the Gulf (Kali, 2003). However, unlike in Talhan, timely intervention by the police brought the Meham conflict under control and the dispute was referred to the court. Presently the Dera is placed under a government receiver who looks after its management.

The conflicts in Talhan and Meham reflect the underlying layers of tensions between the hitherto all powerful and dominant Jutts, and the newly emerged economically independent class of Dalits. Whatever be the causes of conflicts, it is clear that the Dalits in Punjab, especially in the Doaba, have achieved a level of consciousness to assert their rights. In contrast, the Sikh Jutts who have thrived in the meek silence of Dalits, are finding it difficult to grapple with the surging Dalit consciousness. Given the rising level of social consciousness amongst them, the dominant castes are finding it difficult, if not impossible, to ignore their demands for a share in the structures of power at the local level.

CONCLUSION

This paper argues that contrary to the popular view of a casteless Sikh society in Punjab, caste discriminations are very much a part of its social set up. However, what distinguishes it from other parts of India is its indifference to the purity-pollution syndrome. Instead, socio-economic factors and preponderance of 'Jutt cultural patterns' are a fundamental cause of discrimination and oppression of Dalits here.

In Punjab, Sikh Jutts are dominant because of their hold over land and their numerical prepondrance in Sikh religion coupled with their martial nature. Their domination, however, is not rooted in the graded system of caste hierarchy. Dalits for various historical reasons were deprived of land, and their entry into Sikhism could not relieve them of the taint of their lower status. Their landlessness, obviously, made them subservient to the landowning castes, the Sikh Jutts. However, the improved economic condition of the Dalits coupled with their rising social and political consciousness has led to a sharpening of contradictions between them and the Jutts, especially in the Doaba. In fact, Punjab is in a volatile situation wherein the Jutts and Dalits have entangled themselves in a whirlpool of old mindset *versus* rising social consciousness. This in turn has resulted in a series of violent Jutt–Dalit clashes in the state. What weaves the Jutt–Dalit conflicts together despite differences in issues and locations are the similarities in their nature and patterns of emergence. These invariably involve demands of Dalits for a respectable social space in the structures of power at the village level commensurate to their improved economic conditions. Such moves of the marginals find staunch critics among the Jutts who view Dalit assertion as a challenge to their dominant status in rural society.

Despite the fact that agriculture has ceased to exist as a profitable occupation for the last few years, land is still considered the most essential status symbol in rural Punjab. Though many Dalits have benefited from constitutional affirmative action, spread of education, social welfare measures and ventures abroad, a vast majority of them still are landless, very poor and vulnerable. While many

Dalits have abandoned their caste-based occupations and have also distanced themselves from employment in agriculture, their social status has remained marginal, precisely because of their being landless. But it is a fact, that Dalits have achieved significant awareness and political consciousness over the last seven decades in the history of Dalit mobilization in Punjab.

They cannot be coerced any more to remain confined to the periphery. The contradiction between the old mindset based on proclivities of caste prestige and honour, and the emerging Dalit consciousness for equal share in the power structures of the rural society is fast becoming a faultline between the Jutts and the Dalits. The ever-increasing number of caste conflicts is a clear testimony to the emerging dissensions between them. Given the intensity of this consciousness on the part of Dalits, the Jutts are finding it difficult, if not impossible, to ignore Dalit demands without resorting to pressure tactics or force. This, in turn, often leads to clashes between them. A manifestation of Dalit assertion, these clashes have sharpened the issue of their human rights and emboldened the downtrodden to actively engage themselves in the political processes for the realization of their rights.

NOTES

1. Ad Dharm movement came into existence in 1925. It aimed at emancipation of Dalits and their empowerment through cultural transformation, spiritual regeneration and political assertion. It was the first movement of its kind in north India that brought together the downtrodden to fight for their cause. It laid the foundation of Dalit consciousness and assertion in Punjab. Mark Juergensmeyer's seminal work is the pioneer study of this movement (Juergensmeyer, 1988; Ram, 2004).
2. Since the publication of this study many more Ravidass Deras have been established in the state. In 2005 alone, the saints of Ballan have laid the foundation stones of 12 Ravidass Deras (calculated from the *Begumpura Shaher* [Jalandhar] weekly).
3. The Ad Dharm movement helped forge unity among different Dalit castes in the state by bringing them together into the fold of Ad Dharm (an ancient and indigenous religion of the natives of India). This movement

specifically focuses on the ethnification of Dalit identity in the region than on treading the path of Sanskritization to move up the caste hierarchy, as was the case with the Adi Hindu movement (Jaffrelot, 2003; Chandra, 1999). The Ravidass Deras provided the Dalits of Punjab the much-needed cultural space to connect them to their lost cultural heritage. These Deras also provided them the bare minimum infrastructure required for the ethnification of their newly conceived Dalit cultural space. All these efforts helped significantly in the generation of Dalit consciousness in Punjab.

4. However, the main concern of these movements was to transform the attitudes of the individuals rather than striking hard on the asymmetrical structures of society (Grewal, 1994). The socio-religious movements had never taken up the issue of disproportionate landholdings, a crucial cause of social inequality and economic deprivations of Dalits in the state. Whatever small impact the saints and the socio-religious movements were able to bring in the minds of the people faded away with the passage of time.
5. Social boycott, a form of social exclusion, involves a ban on the entry of Dalits into the fields of Jutts. It also involves severe deprivations for the landless Dalits who are dependent on the lands of the Jutts for fuel, fodder and even to answer the call of nature. The Jutt landowners imposed social boycott during the wheat harvesting seasons in the early 1970s, as a weapon of suppression against the landless agricultural labourers who demanded hike in their wages. Nowadays, it is being used against agitating Dalits who ask for equal participation in the formal and informal institutions of power at the local level. According to Judge: 'It is the means to remind them that despite their improved conditions, they continue to be low caste' (Judge, 2006: 12).
6. The rise of militancy in Sikhism in the sixteenth century was generally attributed to the martial nature of Jutts (Habib, 1996: 100; McLeod, 1996: 12). The ranks and leadership of the Khalsa from this period onwards were predominated by the Jutts so much so that the history of the Sikh religion that follows came to be known as 'the history of the Jutt section of the Sikh community' (Pettigrew, 1978: 26). For counter-arguments on this theme see: Singh (ed.), 1986, especially part six; Singh, 1985. In rural Punjab, one often heard a Jutt when suggested to take medicine in case of sickness saying that he would survive even if cut in half.
7. In Islam, Chamars are known as Mochis, and Chuhras are called Musallis and Kutanas. In Christianity Chuhras are named Massihs or Isais. Some of the Chamars who joined Arya Samaj came to be known as 'Chaudhary' and 'Mahashas' (Judge, 2006).
8. Dalits have separate gurdwaras in about 10,000 villages out of a total of

12,780 villages in Punjab (*Dalit Voice*, vol. 22, no. 17, 1–15 September 2003). A survey of 116 villages in one tahsil of Amritsar district showed that Dalits had separate gurdwaras in 68 villages (Puri, 2003). Yet another field-study of 51 villages selected from the three sub-regions of Punjab found that Dalits had separate gurdwaras in as many as 41 villages (Jodhka, 2002; Muktsar, 1999; 2003).

9. Mangoo Ram (1886–1980) was one of the founders of the Ad Dharm movement. Born in a Chamar family in village Mugowal, district Hoshiarpur, Punjab, he immigrated to America (1909) where he came in close contact with the Ghadar Party (a militant nationalist organization). After his return in 1925, he organized the SCs of Punjab against the system of untouchability. During the Round Table Conferences in London (1930–2) he sent telegrams in support of B.R. Ambedkar as the leader of untouchables in India instead of Mahatma Gandhi. In 1946, he was elected to the Punjab Legislative Assembly and remained in legislature till 1952. On 15 August 1972, Prime Minister Indira Gandhi honoured him with a *Tamra Patra* and pension (Rs. 200 per month) for the services he rendered in the Ghadar Party for India's freedom.

REFERENCES

Ahir, D.C. (1992), *Dr. Ambedkar And Punjab*, Delhi: B.R. Publications.

Ambedkar, B.R. (1995), *Annihilation of Caste*, 3rd edn., Jalandhar: Bheem Patrika.

Bakshi, R.P. (n.d.), *Mera Jiwan Sangharsh* (Punjabi), Punjab Pradesh Balmik Sabha.

Chandra, Kanchan (1999), 'Post-Congress Politics in Uttar Pradesh: The Ethnification of the Party System and its Consequences', in Ramashray Roy and Paul Wallace (eds.), *Indian Politics and the 1998 Election: Regionalism, Hindutva, and State Politics*, New Delhi: Sage.

Chumber, C.L. (1986), 'Interview with the Editor of *Adi-Danka*' (Punjabi), *Kaumi Udarian*, vol. 1, no. 2, January, pp. 49–52.

Deep, Dalip Singh (2001), *Sadhan Main Ravidass Sant: Jiwan Ate Vichar* (Sant Ravidass Among the Saints: Life and Thought), Patiala: Punjabi University.

Desh Sevak (Panjabi daily), Chandigarh, 3 June 2007.

Gosal, R.P. S. (2004), 'Distribution and Relative Concentration of Scheduled Caste Population in Punjab', in Harish K. Puri (ed.), *Dalits in Regional Context*, New Delhi: Rawat.

Grewal, J.S. (2006), *The Sikhs of Punjab*, New Delhi: Foundation Books.

Habib, Irfan (1996), 'Jutts of Punjab and Sind', in Harbans Singh and N. Gerald Barrier (eds.), *Punjab Past and Present: Essays in Honour of Dr Ganda Singh*, Patiala: Punjabi University.

Hindu (National daily, Chennai), 16 December 2005.

Ibbetson, Sir Denzil (1883, rpt. 1970), *Punjab Castes*, Patiala: Languages Department, Punjab.

Jaffrelot, Christophe (2003), *India's Silent Revolution: The Rise of the Low Castes in North India*, Delhi: Permanent Black.

Jodhka, Surinder S. (2006), 'The Problem', *Seminar 567* (Re-Imagining Punjab), November, pp. 12–16.

——— (2003), 'Caste, Occupation Divide Gets Sharper in Punjab: Agrarian Technology has Reshaped Social Relations', *The Tribune*, 18 July.

——— (2002), 'Caste and Untouchability in Rural Punjab', *Economic and Political Weekly*, vol. 37, no. 19, 11–17 May, pp. 1813–23.

Jodhka, Surinder S. and Louis Prakash (2003), 'Caste Tensions in Punjab: Talhan and Beyond', *Economic and Political Weekly*, vol. 38, no. 28, 12–18 July, pp. 2923–6.

Judge, P.S. (2006), 'Dalit Assertion in Punjab: Examining New Trends and Emerging Dilemmas', paper presented in a seminar *Conditions of Marginal Groups in India* organized by the Centre for Social Studies, Surat, 5–6 July.

Juergensmeyer, Mark, (2000), 'Ad Dharm Movement', in Harish K. Puri and Paramjit S. Judge (eds.), *Social and Political Movements: Readings on Punjab*, Jaipur and New Delhi: Rawat.

——— (1988), *Religious Rebels in the Punjab: The Social Vision of Untouchables*, Delhi: Ajanta.

Kali, Des Raj (2003), 'Nurmahal vich vi Talhan bannan di udeek kar riha parshasan' (Administration is Waiting for a Talhan-like Situation to Emerge in Nurmahal), *Rozana Nawan Zamana*, Jalandhar, 29 August.

Kaur, Ravinder (1986), 'Jutt Sikhs: A Question of Identity', *Contributions to Indian Sociology* (n.s.), vol. 20, no. 2, pp. 221–39.

McLeod, W.H. (1996), *The Evolution of Sikh Community*, New Delhi: Oxford University Press.

——— (1986), *Punjabis in New Zealand: A History of Punjab Migration 1890–1940*, Amritsar: Guru Nanak Dev University.

Muktsar, Gurnam Singh (2003), 'Sikhs Divided into 3 Warring Camps: Upper Castes, Jutts and Dalits', *Dalit Voice*, 1–15 August , pp. 21–2.

——— (1999), 'Every Jutt has Separate Gurdwara Even as Sikhs are Busy with Khalsa Fete', *Dalit Voice*, 16–30 June.

Palshikar, Suhas (2004), 'Revisiting State Level Politics', *Economic and Political Weekly*, vol. 39, nos. 14–15, pp. 1477–80.

Pawar, I.D. (1993), *My Struggle in Life*, 3rd edn., Chandigarh: Author Publishers.

Pettigrew, Joyce (1978), *Robber Noblemen: A Study of the Political System of the Sikh Jutts*, New Delhi: Ambika Publications.

Philip, A.J. (2003), 'All in the Name of Almighty: The Caste Divide at Talhan', *The Tribune*, 16 June.

Piplanwala, Babu Hazara Ram (1986), 'Lothian Committee and Ad Dharm Mandal', *Kaumi Udarian*, vol. 1, no. 2, January, pp. 10–15.

Puri, Harish K. (2004), 'Introduction', *Dalits in Regional Context*, Jaipur and New Delhi: Rawat.

——— (2004a), 'Dalits of Punjab', *International Newsletter* (Waterford, CT), vol. 9, no. 1, February, pp.1, 9–11.

——— (2003), 'Scheduled Castes in Sikh Community: A Historical Perspective', *Economic and Political Weekly*, vol. 38, no. 26, 28 June–4 July, pp. 2693–2701.

Rajshekar, V.T. (2004), 'A Silent Socio-cultural Revolution Sweeps Punjab: Ravidass Saints Gain Millions of Followers', *Dalit Voice* (Bangalore), vol. 23, no. 13, 1–15 July, pp. 3–5.

Ram, Ronki (2004), 'Untouchability, Dalit Consciousness, and the Ad Dharm Movement in Punjab', *Contributions to Indian Sociology* (n.s.), vol. 38, no. 3, September–December, pp. 323–49.

——— (2004a) 'The Dalit Sikhs', *Dalit International Newsletter*, vol. 9, no. 3, October, pp. 5–7.

——— (2004b) 'Untouchability in India with a Difference: Ad Dharm, Dalit Assertion, and Caste Conflicts in Punjab', *Asian Survey*, vol. XLIV, no. 6, November/December, pp. 895–912.

Saberwal, Satish (1976), *Mobile Men: Limits to Social Change in Urban Punjab*, New Delhi: Vikas.

——— (1973), 'Receding Pollution: Intercaste Relations in Urban Punjab', *Sociological Bulletin*, vol. 22, no. 3, September, pp. 234–59.

Sain, Chattar (1985), 'Ghadri Ton Ad Dharmi: Babu Mangoo Ram Ji Mugowalia' (Punjabi), *Souvenir*, Babu Mangoo Ram Mugowalia 99th Birth Anniversary, New Delhi, pp. 35–7.

Sharma, Reeta (2003), 'Flames of Caste', *The Tribune*, 21 June.

Singh, Balwant (2002), *The Suffering People*, Saharanpur: Ambedkar Mission Publication.

Singh, Gurdev (ed.) (1986), *Perspectives on the Sikh Tradition*, Patiala: Siddharth Publications.

Singh, Jagjit (1985), *Perspectives on Sikh Studies*, New Delhi: Guru Nanak Foundation.

Singh, Khushwant (1953), *The Sikhs*, London: George Allen & Unwin.

Singh, Inder Pal (1977), 'Caste in a Sikh Village', in Harjinder Singh (ed.), *Caste Among Non-Hindus in India*, Delhi: National Publishing House.

Singh, Manjit (2005), 'Introductory Remarks', circulated at the National Seminar on Future of Rural Development in the North West India, organized by the Department of Sociology, Panjab University, Chandigarh, 24–5 February.

——— (2001), 'Economic Change and Dalit Identity in Rural Punjab: A Note', paper presented at the seminar on *Liberalisation, Dalits and the State*, organized by the Department of Sociology, Panjab University, Chandigarh, 3–4 May.

Singh, Prabhjot (2003), ' "Mazaar" That Sparked Violence', *The Tribune*, 10 June.

Tandon, P. (1961), *Punjabi Century (1857–1947)*, New York: Harcourt, Brace and World.

Thorat, Sukhadeo (2006), 'Paying the Social Debt', *Economic and Political Weekly*, vol. XLI, no. 24, 17–23 June, pp. 2432–5.

Verma, Archana B. (2002), *The Making of Little Punjab in Canada: Patterns of Immigration*, New Delhi: Sage.

Virdi, S.L. (2003), 'Bharat De Dalitan Di Asseem Dastan Da Dastavej' (Punjabi), *Begum Pura Shaher*, 11 August, pp. 2–11.

——— (2003a), 'Punjab'ch Zimindaran Atey Dalitan Vichkar Takrar Dalit Chetna Da Prateek' (Punjabi), *Begum Pura Shaher*, 18 August, pp. 6–7.

Yadav, Y. (1999), 'Electoral Politics in the Time of Change: India's Third Electoral System, 1989–9', *Economic and Political Weekly*, vol. 35, nos. 34–5, 21–7 August and 28 August–3 September, pp. 2393–9.

CHAPTER 13

Des Hoya Pardes: The Crisis of Punjabi Peasantry and Sikh Emigration Since the 1970s

Darshan Singh Tatla

INTRODUCTION

Almost everyday vernacular newspapers in Punjab report scandals involving travel agents duping would-be migrants. So when the *Hindustan Times* carried a two-page supplement on the aftermath of the Malta boat tragedy, which carried interviews with several affected families, a young man whose father had died in the tragedy telling the reporter how 'he is determined to go abroad' did not attract any editorial comment or a campaign against unscrupulous travel agents.[1] Why was this young man so determined to leave his home? Significantly, he was and is not alone. For almost a decade now such news and views have consistently appeared in the media. Numerous advertisements appear in Punjabi newspapers enticing readers with various packages of overseas immigration for temporary work and permanent settlement. There are special magazines devoted exclusively to news of overseas opportunities along with features on successful diaspora Punjabis. Travel agents have over the years acquired experience and specialized ways of smuggling Punjabi youth into the Middle East, then facilitating their passage to eastern Europe. For further fees they can arrange migrants' entry into western European countries. For such hazardous and illegal human trafficking, the costs vary from anything between a quarter to half a million rupees for the Middle East and over two million rupees for the ultimate destination, the United States. There are numerous instances of young men having been flown to a different country than their promised destination or dumped somewhere at midnight without warning. There are many others who have lost all their money on the way and unable to gain any promised employment. Reports cite how such youth eventually end up in prisons and slowly trek back from European prisons through the Middle East into Pakistan and after much haggling and appeals to authorities, return home, their unsuccessful sojourn lasting several years.[2]

Why are Punjabis so desperate to leave their homes? It is surprising that such families neither face famine nor are below the so-called poverty line. Then, is it something to do with personal perception of their economic and social position within the society or is it a phenomenon related to the Sikhs' mass psychology? Moreover, as anyone familiar with the contemporary Punjab knows, emigration is not an isolated phenomenon confined to a particular class of people or to a particular region. Migration has emerged as an industry in Punjab involving thousands of travel agents, transporters, hoteliers, government officials and non-Indian middlemen.[3]

The question naturally arises, why is there so much pressure to emigrate from Punjab, especially among its rural population and mainly the Sikhs? This paper seeks to highlight some crucial changes in the nature of Punjabi emigration in the last three decades and offer an alternative explanation for such an exodus. It is suggested that the rising pace of Sikh emigration and the emergence of a 'culture of migration' has been facilitated by the peculiar impact of the post-colonial Indian state's institutions and policies towards its peasantry. That is to say that the increasing propensity among the Sikh peasantry to emigrate has to be located in its characteristic experience of modernization since the 1970s. It is argued in the following pages that the rising standards of living accompanying the success of the Green Revolution was tinged with a collective communal experience of 'despair', 'dishonour' and 'death', the three concepts that are explained in this paper. It is contended that 'governance', that is, the role of state institutions, and the post-colonial Indian statecraft as perceived by the peasantry emerge as crucial factor in explaining the rising tempo and diverse destinations of current Sikh emigration process.

CHARACTERISTICS OF SIKH MIGRATION SINCE THE 1970s

A cursory examination of contemporary emigration pattern from Punjab alerts observers to how almost all regions of the province as well various rural classes are affected by it. This is a major departure from early phases involving small number of people from certain specific areas of the province. Although there is paucity of research on migration in general and emigration process in particular, certain characteristics of Punjabi emigration process can be discerned since the 1970s when its tempo started to accelerate.

The emigration process which coincided with the period of Green Revolution seems paradoxical at first. The rural society as a whole was reaping rewards of increasing output of wheat and the introduction of rice. High-Yielding Varieties (HYV) were augmenting farmers' incomes as also those of farm labourers. At the same time, an influx of labourers from Bihar and western Uttar Pradesh was noticeable as local labour was stretched to the limit for sowing and harvesting of rice and wheat and to meet the increased labour demand for farming. This was also the period when Punjabi farmers and labourers were emigrating abroad in increasing numbers. Indeed the dual process of in-migration and out-migration were taking place simultaneously.

This emigration process was also spreading to regions which were quite immune to it in the early era. Thus, districts of Patiala and Ropar in the east joined others. Perhaps the only anomaly was the Majha region which had earlier been well connected abroad, but was now left behind. This rather peculiar aspect needs further analysis, but it is a matter beyond the scope of this paper. In terms of classes and segments of Sikh rural society, the emigration fervour has embraced them all. Even prosperous farmers appear as eager to send their off-springs abroad as are the middle and small land-holders. One hears desperate stories of emigration from all the classes. In Doaba, Majha and Malwa regions, many peasant families have sold their land, in some cases the entire family land, to finance a son's or a daughter's settlement abroad.

Thus current emigration is not confined to just one class but encompasses a broad cross-section of peasantry. One has to depend upon general impressions in the absence of empirical data, but it seems all castes/classes are equally scrambling for opportunities abroad. Many of erstwhile maharajas' descendants, one discovers, have either settled away from Punjab or actually left for abroad. If one examines the class of civil servants or the salaried elite, that is, personnel working in higher education, major professions such as law, health and social services, and so on, one notices that they too have joined the emigration process.[4] One should also note this trend amongst the political leaders as almost all of them have a foot abroad. Lok Bhalai Party chief Balwant Singh Ramoowalia has all his immediate family settled in Canada; appropriately he has championed the cause of overseas Punjabis to earn his popularity for whatever its worth. In the past he has taken up the issue of abandoned brides by overseas grooms, and harsh treatment meted out to Punjabi visitors at Delhi and Amritsar airports. All leading activists of major political parties—the Congress, the Akali Dal and the Communists—are well-connected abroad through their sons, daughters or other kin.

The means used for emigration are as diverse as the destination and routes taken by the migrants. Travel agents can smuggle prospective immigrants into the Middle East, western Europe or North America; failing that they compromise with some destination in the Far East. In the last decade, Australia and New Zealand have come on the 'market' where youth obtain admissions through 'on-the-spot-interviews' conducted either by the representatives of universities or colleges from there or their franchises and Indian representatives at various city hotels in the major cities of Punjab. Indeed, there is a whole class of qualified young men fleeing abroad. An Australian city is reputed to have a 'Ph.D. Taxi Stand' maintained by Punjabis (Voigt–Graf, 2003). Visas are obtained relatively easily for educational diplomas, degrees or vocational training courses. The dubious representatives promise 'guaranteed' visas and charge hefty fees. They claim that such courses are certified by prestigious institutions abroad. Such claims often turn out to be lies. In a nutshell, the Punjabi youth are desperately seeking overseas opportunities on their own or through unscrupulous agents offering various packages.

Several socially reprehensible practices have emerged in the

Punjabi society showing their intense ambition for going abroad. One is 'fake marriage' as an avenue for migration in which the relatives often sponsor one of their kin through such an alliance while unrelated families reciprocate 'convenient marriage' facilitating two separate families to migrate. Cases are also known of foreign-settled single men and women, divorced or even already married, on their own or through marriage bureaus/agencies, charging fees for 'fake marriage', duping not one but many aspirant migrants simultaneously. Indeed, 'marriage racketeering' is now a part of rural Sikhs social scene. During the winter months, many agencies offer all-under-one-roof package for such a marriage, providing the 'groom' or 'bride', along with hired in-laws and suitably clad guests and hosts from each side and a false religious ceremony in the presence of real or a replica of the *Guru Granth Sahib*.[5] In many such cases of dubious marriage transactions, the contracting parties are 'respectable' families willing to undergo such 'dishonourable' deals. The widespread use and acceptance of such malpractices besides underlining a steep demand for overseas settlement, also indicates a radical change in the values of the Sikh peasantry and its evaluation of social hierarchy in which 'emigrant status' has come to play crucial role.

CONVENTIONAL THEORIES

Why is there desperation among Punjabis to emigrate abroad in such large numbers? Why are prosperous families under such an invidious pressure to emigrate? The purely economic motive, that is, to better themselves and their families through emigration, is obviously an unsatisfactory explanation for such a massive flight of people. The economic motive alone is an insufficient variable to explain such a flight of people affecting the rural Punjabi (Sikh) society.

A standard explanation for migration runs in terms of pull/push factors. Here the unit of analysis is usually the household as the decision maker. Even when extended to kinship networks, usually there is no reference to the community's perception of the migration process.[6] Though understandable in terms of empirical verification, however, an analysis based on household units and kinship relations is inadequate in explaining more recent Punjabi migration.

We can still make use of 'pull and push' factors for explaining migration in certain periods. There appears to be a close fit between migration in the pre-1947 period and changes in various countries' immigration rules—of colonies, the dominions and of the United States—this factor by and large determined the fluctuation and overall flow of Punjabis/Sikhs to such destinations. The same pull factors can also account, though a bit more loosely, for the migration pattern up to the 1970s. Thus, overseas governmental rules and regulations affecting Indian subcontinental immigration would have affected the overall numbers and pace of emigration, but it does not explain Punjabis' distinctive response to such rules. The United States adopted a more open policy towards skilled labour from the Indian subcontinent by raising its almost barred zone policy after the bill of 1948. The relaxing of immigration rules for blood relations in the case of North American immigration laws from the 1950s and 1960s were major factors in re-starting Punjabi migration to Canada and USA. Similarly as the UK opened its doors to Indian immigrants from the late 1950s, Punjabi migration especially from the Doaba region increased sharply through the 1960s. Increasing immigration to North America and Britain since the 1960s can be explained in terms of the relaxed policies of these three countries as a very small number of Punjabis already settled there could call their kin. In the case of Canada, it was projected by a historian that the first rush was indeed of the relatives and such was the spiral effect that by the 1970s, a small number of families could account for most of such immigration with an ever increasing number of kin and relatives moving through chain-migration.

Similar pull factors can account for change of direction in the destination of Punjabi migrants. Thus, countries of older settlement such as East Africa and the Far East gradually drew shutters making it increasingly difficult for fresh migration. While Punjabi immigration to Singapore, Malaysia, Hong Kong, and Thailand in the Far East and Kenya, Uganda and Tanzania in East Africa came to an almost standstill, later in the 1990s, Australia and New Zealand started accepting skilled labour drawing many youngsters. Another avenue opened up in the 1970s when several Middle Eastern countries created huge demand for labour as a result of surplus petro-dollars earned following the hike in oil prices. From the Indian subcontinent, Kerala and Punjab became major suppliers of this short-term labour for vast construction projects undertaken by multinational companies in Iraq, United Arab Emirates (UAE), Dubai, Kuwait and other Middle Eastern countries.

Consideration of push factors is usually couched in the economic and social environment conducive to the migrants' flow. For Punjabi/ Sikh emigration, an explanation runs in terms of *pressure on land*. It is well known that Punjab's peasantry has, comparatively speaking, small landholdings. This has meant a 'pressure on land' or what is called 'surplus son syndrome'. All it means is that land is always short for brothers who inherit it from their father. This theory was used to explain army recruitment as a peasant's family consisting of three or more sons was likely to send a son into the army. Certainly, Doaba's predominance in emigration can be partially explained in terms of the scarcity of land as Jalandhar and Hoshiarpur districts have not only smaller landholdings but the soil too is of poor quality. In other words, such migration was of marginal peasants who were driven abroad in a bid to escape from poverty. Many families owning only a little land or no land at all went abroad, but it seems this accounted only for a small proportion of the total number of emigrants. Among the migrants one certainly finds on overwhelming ratio of those who were comparatively better off in terms of the village society. That they chose to emigrate was not to avoid stark poverty or economic compulsion, but to enhance their status *vis-à-vis* other families.

Thus, conventional interplay of push/pull factors or chain migration familiar to sociologists becomes far less useful in accounting for the post-1947 Sikh/Punjabi emigration pattern, especially after the 1970s. For the latter period, we need to account for the collective Sikh behaviour as a major motivating factor. Collective behaviour has been defined as

> forms of social behaviour in which usual conventions cease to guide social action and people collectively transcend, bypass or subvert established institutional patterns and structures . . . collective behaviour refers to the action of collectivities, not to a type of individual behaviour. (Turner and Killian, 1957: 3)

The numbers involved and the pattern of migration evolved during recent decades certainly suggests it to be a mass behaviour where individual decision-making process has become a less determinant factor. In other words, economic benefits and rewards of migration which had encouraged individuals to migrate have transformed into cultural meanings resulting in a 'culture of migration' transcending the economic aspects.[7] The net result is that migration has become part of Punjabi peasantry's normative values, connected to status and as a social mobility, and as a crucial means of economic survival. Going abroad not only ensures economic survival, but enhances prestige and status. Since the 1980s, it has also become an escape route from an oppressive state, its administration and misgovernance. Migration has come to mean a search for social and economic security which Punjab's own institutions have failed to uphold.

Coming close to such an analysis of Sikhs' collective behaviour, Pettigrew suggests that Punjabi migration can be better explained through a more subjective factor, namely *enhancement of status.* Pettigrew finds that Sikh ethics as understood by Jats has meant a widespread prevalence of *competitive material ethics*, that is, in the acquisition of worldly goods and assets, they have competed against each other (Pettigrew, 1971). She also locates a Jat Sikh's decision to emigrate to seek freedom from a culture of 'suffocating obligations and ties' to one's kin. While Pettigrew comes close to addressing the 'collective decision-making process affecting the Sikhs', however, internal social structure is just one aspect. What we need is another missing variable which could offer further explanation for the collective behaviour of the peasantry as an 'emigrating society'.

EXPLAINING SIKH MIGRATION: 'GOVERNANCE' AS A FACTOR

The question arises why so many Punjabis, especially Sikhs—the main focus of our paper—are leaving Punjab? Why should the Sikh peasantry assign so much value to migration, treating it almost a panacea for all of life's ills? In a word, why has Punjab become such a disappointment to the Sikhs? This has been imaginatively captured in a recent film *Des Hoya Pardes*, literally translated as 'Homeland Turns into a Foreign Territory'.[8] We have already argued that the post-Green Revolution Sikh migration has to be treated as a collective behaviour, so the question arises what are the main factors contributing towards such a behaviour? Briefly, the answer is that such collective migrating behaviour is to be located in the crisis of Sikh peasantry in the aftermath of the Green Revolution. The paper argues that while falling returns to agricultural production after the initial success of the Green Revolution was one factor giving rise to frustrations and disappointments, the other crucial factor was Punjabi peasants' gradual and collective perception of 'the state as an enemy'. Thus they employed all possible means to flee from its clutches. While this is elaborated below, the paper contends that conventional theories of migration need to be supplemented by taking cognizance of the socio-political framework of Punjab's political economy in post-colonial India.

MODERNIZATION EXPERIENCE OF THE SIKH PEASANTRY

Agrarian change in Punjab since the 1970s has been the subject of several studies (Shiva, 1991; Jodhka, 1997, 2001; Iyer and Manick, 2000; Gill and Ghuman, 2000; Shergill, 1998; Kumar and Sharma, 1998). Most of these studies point towards the peasantry suffering from a crisis through a gradual reduction of profit margins on agricultural production. These studies are almost at a consensus that after the initial success of the Green Revolution when the Punjabi peasantry had a brief spell of economic prosperity, the process began to reverse as small peasants with landholdings of 5 acres or less felt squeezed by the very success of this Revolution. Punjab peasantry was by no means ill-prepared for the transformation to be caused by the Green Revolution. Many of the peasants, especially those who returned from Western *bar* areas, were experienced farmers used to many kinds of innovation in crop production. Within a short span of time they had been instrumental in changing the ecology and agrarian structure of a vast stretch of land. They brought skills as also a risk-taking attitude, with them and this was one crucial factor in the successful implementation of the intensive agricultural strategy in Punjab.

Those at the lower end of landholdings were unable to cultivate their lands due to spiralling costs of agricultural input while support prices for two major crops, rice and wheat, did not rise proportionately. It is also noted by some economists that generally there were worsening terms of trade against agriculture which had added to the farmers' woes. Peasants' dependence on rice-wheat cycle of crops for their incomes led to sharp reduction in their economic fortunes as prices of their output was determined by the central government's fiat.

The increasing cost of pumping of ground water, which has gone deeper and deeper, has also added to the woes of the peasantry. Most farmers have to spend on deepening their pump wells almost every two years. The water requirements for rice far-exceed the ground water resources in the state. Canal water is also insufficient for rice crop. Thus water has indeed become the basis of much contention among individual farmers as also in collective political mobilization.

Having to sell in the market at prices determined by the central government has contributed to the peasants' sense of loss over their own affairs. Increasingly, they feel frustrated in their economic endeavours, while the crisis literally turned into a grave for many small peasants as is clear from a series of suicides by peasants since the 1990s.[9] More recently, the villagers have offered their whole village for sale to service their collective debt. Besides declining returns for individual farmers, the collateral damage to Punjab's ecology and environment through fixed wheat-rice production has been profound though little studied. Remarkably, this situation in the very first decade of the new millennium compares with the same decade of the twentieth century when farmers' indebtedness gave rise to serious discontent. The government then had responded with tough legislation on the transfer of lands (Barrier, 1964).

In a word, the modernization process has led to a sense of collective despair. This is seen in the pauperization of small landholding peasantry, an increasing tendency in them to lease or sell their lands and to commit suicide due to indebtedness. This holds true all across the state. Modernization has also led to the politicization of the peasantry, with the issue of support prices for rice and wheat becoming crucial catalysts of periodic mobilization. Punjabi peasantry's modernization experience attending the Green Revolution was into transformation from traditional rural folks to aggressively politicized and commercial class of *kulaks*. This modernization process has changed the outlook of the peasantry; its social world has disintegrated; content has been replaced by greed for material goods, and its expectations from the government and its institutions have been inflated. After an initial spell of economic prosperity, the overall experience is one of cumulative frustrations and 'despair' especially among the lower and middle-class Sikh peasantry.

Economic failure has been accompanied by mental disarray reinforced by subsequent political and social developments in the province. The diminishing fortunes in agriculture have brought several social conventions into disrepute. Although the notion of *izzat* or honour has somewhat dimmed since the days Darling wrote of 'the most precious ornament of Jat Sikhs which can lead easily to murder', it still matters dearly to the Sikh *biradri*. Socially, the modernization process has progressively disintegrated the old ties and kinship with its supportive structure but without replacing it with an alternative means to sustain it in financial and social crisis. It is noted that the

> . . . role of traditional institutions like the *biradari*, the village panchayat, religious and other humanitarian institutions which were providing mutual help and social security are now under great stress. . . . The economic development . . . dominance of the capitalist value system, emergence of consumerist culture and the role of the mass media have contributed jointly to the degeneration of traditional value system and is overtaken by the value of individualism. (Iyer and Manick, 2000: 101)

Culturally too, the new socio-cultural milieu seems to have alienated many peasants as they are unable to identify with its various manifestations in visual and expressive arts, and this is seen through their passive disapproval of the television and other media. This despite the already established transnational social networks that link the Punjabi Sikhs to various host countries that have played their role in transforming the tempo of migratory flows into a self-sustaining migration system (Boyd, 1989; Snowdon, 1990). Over time such socio-cultural factors have reduced the importance of economic conditions as the primary causal force that motivates the Sikh peasantry for migration. Rural hierarchy based on landholdings underwent severe changes in the aftermath of the Green Revolution years. Power and prestige of land was dwindling against rich pickings through migrants' remittances, and ostensibly displayed through modern houses on the outskirts of ancestral villages. Alternative to land appeared—send a son abroad. It became an acceptable endeavour, an effort to form the 'honourable fraternity' of the village. Thus, for example, having to sell land for a daughter's marriage is now almost a necessity for marginal farmers. In earlier times it was unthinkable to sell land outright. The usual course was to borrow money from the local *bania*, mortgage land to him and pay him back over the years. But financing a passage abroad for a young man of the family by selling an acre or two became an acceptable option. If a daughter could be married to a Non-Resident Punjabi (NRP), as seems to have become a distinct possibility during the last decade, then selling land was fully justified as in a few years, the whole family could potentially go abroad. Among the peasantry who face formidable challenge to service their debt from ever decreasing returns, many youth have taken to foreign shores after selling the last acre of land, incurring further debts for the passage.

MIGRATION AND POLITICAL CRISIS

The economic crisis of the Sikh peasantry alone could not force migration as a collective behaviour. At best it could add to an already well-established chain-migration from Punjab where individuals and families who have previous kin connections or the means go abroad. What turned migration into a collective endeavour and almost a panacea for all ills of peasantry is a post-colonial Indian political experience.

It is unnecessary to outline the origins of this movement started in the 1970s by the Akali Dal, seen by many as the political party comprising and representing the interests of Jat Sikh peasantry. With much of its membership consisting of Sikh peasantry, the Akali Dal since its inception in the 1920 has claimed more than just material interests of the Sikh peasantry. It has claimed to be a champion of Sikh culture, language and religious tradition, indeed acting on and articulating the demands of Sikhs' political status ranging from seeking reservations to a separate Sikh homeland. In independent India, the Akali Dal has resisted Indian state's integrationist policies, claiming at first special status in Punjab as its homeland. Failing that it launched several mobilizations towards a Punjabi-speaking state. This campaign lasting more than 15 years ended with a smallest possible geographical Punjab based on a 'disputed and controversial linguistic criterion' along with Chandigarh—a city constructed as its capital, becoming the shared capital of two provinces, Punjab and Haryana, created in 1966. What the newly demarcated Punjab achieved was to make Sikhs a majority community in a state for the first time in its history, and the Akali Dal became a ruling party through coalition with others, ousting the Congress party. Its power assumption at the provincial level gave the peasantry a fleeting satisfaction of having become masters of their destiny where the state would safeguard not only their economic interests but perhaps promote its religious values and Punjabi culture too. However, such optimism was short-lived and got immediately thwarted by Punjab's demography, compulsions of electoral politics and the power structure of the central government.

In the late 1970s and early 1980s, a campaign was launched by the Akali Dal with specific and limited demands for an independent, sovereign Punjab. These demands based on the *Anandpur Sahib Resolution* were to end in a devastating tragedy in June 1984 when the Indian government responded by ordering the armed forces into the Golden Temple, destroying its second most sacred building, the Akal Takht, along with extensive damage to the whole of the sacred complex. The crisis of Sikh peasantry which began with declining returns on agrarian produce was laid bare by the central government as it decided to crush any illusion of the peasantry's political autonomy or control over their own affairs at the provincial level. As a reaction to this event, there began a massive resistance and revenge movement which lasted over a decade and claimed the lives over 80,000 Sikhs, mostly youth, and over 20,000 security personnel, mostly of the police. Indeed the indiscriminate use of the state's might was such as to fear a general genocide of a religious minority. Moreover, if the Akali campaign was to register the Sikh peasantry's economic, social and cultural crisis, the federal government's action in sending the army into Punjab crushed any illusion of the peasantry's political autonomy or control over their own affairs at the provincial level. As Pettigrew observed:

> . . . immediate causes that have turned a farmers' movement initially concerned with socio-economic issues into national struggle for freedom. Issues of religious identity and economic discrimination important as they were, eventually became mere dimensions of the struggle as the Sikh as a people faced the elimination of their very young men. (Pettigrew, 1995: 4)

What is important to note is that ordinary people caught between the militants' and security forces' terror had to suffer several kinds of indignities; cases of rapes and brutal killings in real and staged encounters became routine. Families or individuals who fell victims to the state's or militants' repression saw no end to their troubles. The fate of known militant family members was intolerably bad. When a militant was killed, the police or paramilitary authorities usually picked another relative until the family lost support within the village and felt 'dishonoured' beyond grief.

The Akali leaders, mostly drawn from the peasantry, had to swallow the plain fact that the Indian state structure commanded nothing less than submission and cooption with it. The net result was that a community's integrity and the will of its members to remain as 'proud Sikhs' was shattered by the decade-long state repression. For several members of the community, the state effectively posed as an 'enemy', a massive repressive machinery caring little for the security and welfare of its 'own citizens'. The central government treated the demands presented to it as no more than law and order issues:

The state framed an instrumentalist and a mechanical vision of the Punjab problem, a problem of law and order and took measures to settle that primarily by strengthening and modernising the repressive state apparatus. (Singh, 2002: 119)

What could a family do to mitigate its 'honour'? In many cases, it would put its resources together to smuggle a family member abroad, or use any overseas connection to facilitate such emigration. With 'death stalking the countryside' many families which perceived threat of authorities arranged to send their sons abroad. In some cases, it meant making an exaggerated claim of police prosecution that was thought fair in view of the prevailing atmosphere of terror and confusion. If emigration was not a viable option, then migration from the village to a more anonymous and safe sanctuary of a nearby town or city was undertaken though it was hardly a substitute for overseas escape which could rehabilitate their honour besides promise economic relief.

Indeed, it is possible to see how the capitalist development in Punjab's economy led to one of the most tragic and violent movements against the state in the 1980s. As several economists and political scientists have contended, the resistance to state can be seen as Jat-Sikh peasantry's dissatisfaction with the Indian state's policies. Gradually, during this period, most peasants had a steep learning curve about the Indian statecraft as also their own provincial political leaders, especially the Akalis. Or, at least, this can be applied to those exposed to the media and the politicized section of the peasantry. They learnt that before the absolute power vested in the centralized Indian state, the provincial legislatures and their representatives were mostly irrelevant to the decision-making process. That major decisions regarding Punjab's economy and society are taken by the central government and the provincial government is like a mere municipality with paltry powers of appointments to village and town level administration. That neither could Punjab decide about the price of its produce, nor undertake to implement any industrial policy without a nod from the central government. In the much debated issue of agrarian diversification which currently occupies the provincial government led by the Congress party chief minister, Amarinder Singh, the province cannot break the vicious circle of dependency on rice–wheat without financial assistance from the central government. Although it is suggested that globalization has somewhat eased the strangulating grip of the central government on provincial initiatives, all major investment policy decisions still rest with New Delhi. Even the exploitation of overseas Punjabis' resources which could substantially augment Punjab's finances, need suitable legislation and regulations from the central government which has different priorities in attracting the Non-Resident Indians (NRI). Finally, many regulations and laws passed by the provincial government usually end up being questioned or quashed by the Supreme Court or by the central government's overriding instructions. Thus the Sikh peasantry has realized that the provincial government has not only mismanaged the economy, but also that has neither the capacity nor the constitutional power to effectively take charge of the leading sectors of the provincial economy. Thus, as far as Punjab's provincial state apparatus is concerned it has been effectively reduced to an important spectator shorn of governing powers and rules.

It is in this context that massive migration from Punjab should be seen as a rational reaction of a peasantry in crisis. Though disenchanted with their homeland, the land of their dreams, they remain attached to it. Despite abandoning their homeland in increasing numbers, their affection for the land has not decreased. This may be seen in the numerous social, educational and cultural projects sponsored and financed by the Sikh diaspora in their respective villages or towns.[10]

That the state can be seen differently by different sections of society and that its policies have differential impact on different sections of society is not a new revelation in the social sciences. What is new is that political scientists are now increasingly concerned about the nature of state power. In a federal polity such as India, state power can often be exercised to exclude a segment of population from normal participatory democracy or even subject it to repression to achieve ends that might not find consensus among minorities or certain targeted groups. Moreover, a state which governs a diverse population and territory with ethnic homelands could also act as a democratic institution for the majority community, while for a particular minority its policy and action may tantamount to that of a 'criminal or rogue state'. Thus the Indian state and its authority in Punjab during the 1980s were perceived by a majority of Sikhs as unrepresentative of their will. The state power was manifestly held by unrepresentative politicians protected by gun-trotting security personnel in high-speed Gypsies with red-blinkers resembling the popular image of broad day-robbers.[11] Indeed as militants gained hand in certain parts of Punjab in the late 1980s, it was a failing state which became 'criminalized' with Draconian measures of 'bribing disheartened security and police forces to fight' on its behalf (Hetchter, 1987; Cliffe and Luckham, 1999; Green and Ward, 2004).

While the Sikh masses and its politicians were to learn the lessons of 1980s under duress, there were other after-effects of the resistance movement that have continued to shadow Punjabi lives. As a semblance of normalcy prevailed from the mid-1990s, the cost of fighting 'terrorism' became apparent. An increasing share of revenue expenditure was devoted to fighting militancy, thus eroding the fiscal stability of the state finances. The situation was worsened by the fact that the state government could not administer an efficient tax regime. According to a report by the World Bank on Punjab economy, from 1980 to 2000 there was a sharp economic decline (World Bank, 2004). When the state was under President's Rule during 1987–92, Punjab's civil services grew by 3 per cent per annum and the total governmental staff stood at 3,76,222 in 2002.[12] The police budget doubled to 8 per cent, consuming Rs. 1,000 crore of public money and the numerical strength of its personnel expanded exponentially.[13] The report also pointed out how corruption played a part in recruitment, transfers, tendering and procurement and service delivery.[14]

The state has failed even more strikingly in its social welfare provision. Health and education sectors declined sharply in quality in the last two decades, i.e. 1980s and 1990s. Almost 94 per cent of the health department's budget was spent on salaries and pensions to its employees, leaving just a pittance for maintenance, supplies and materials including drugs in government hospitals and dispensaries. The state of education was worse with 98.5 per cent of the sector's allocation spent on employees, while quality of education declined.[15] In both health and education sectors, unchecked entry of private service providers affected the morale of public sector employees while unplanned expansion of liberal and technical education added to an already large pool of the 'educated unemployed'.

Thus, the search for overseas opportunities by Punjabi youth, especially from the rural Sikh peasantry, cannot be delinked from the state's failure in providing gainful employment to its qualified and educated citizens. With the mushrooning of educational institutions, it is not surprising that there are thousands of chronically under-employed skilled men and women in Punjab.[16] This oversupply of educated men and women has been the direct result of unplanned expansion of higher education in the province.[17]

At the peak of militancy during 1987–92 many male members of the affected families from Majha, Malwa and Doaba regions of Punjab escaped from the country, some taking the hazardous routes of Nepal and Pakistan. Some of them could rely on connections within the Sikh diaspora who provided initial support for such moves. A few militant leaders were sponsored by overseas Sikh organizations who also vouched for their 'refugee' status. During this period, thousands of Sikh youth landed at airports in the Far East, Western Europe and North America. Once allowed entry, they could depend upon sympathizers and supportive organizations for their claims as refugee settlers.[18] The World Sikh Organization, the Babbar Khalsa International and International Sikh Youth Federation and several other organizations have provided notable help.[19] But claims for refugee status have found stiff contests from the Indian embassies. In any case, the decision on their application for asylum meant they could stay for several years and acquire financial security even if they were eventually deported. The Indian state concluded extradition treaties with countries of notable Sikh populations. The first such treaty was signed with Canada in 1985, prompted by the crash of Air India flight 182 'allegedly a Canadian Sikh's conspiracy', killing all of its 329 passengers on board (Mukherjee and Blaise, 1988; Tatla, 2004).

SIKH ANOMIE?

Although relative normalcy returned to Punjab in the mid-1990s, the defeat of Sikh militants by the Indian security forces had a profound demoralizing effect on the whole community. Although participation in public life has begun again, the Sikh peasantry has since acquired a resigned outlook of despair along with feelings of betrayal. Possibly psychologists could dissect the fractured inner world of the 'Sikh persona' which gradually came to replace the former self-confident assertion of communal pride. The 'integrative' Sikh consciousness seems to have broken down into a self-destructive competitive ethos of seeking power and goods through fair and foul means. In particular, the peasantry's ethical value system stands dissolved both as a result of modernization as also due to indignities suffered in the aftermath of the 1984 events, giving way, it transpires, to one overwhelming resolution—leave the ill-governed Punjab.

From the 1970s onwards, leading institutions of communal trust have been compromised through the dual process of modernization and state interference. The rapidly enforced transformation of Punjab's agriculture has resulted in profound ecological crisis, while the substitution of farming as a way of life into a semi-capitalist system of 'cash produce for the market' has led to severe adjustments of lifestyle and thinking process by the peasantry. From a close-knit community of small neighbourly communities having 'organic' relationships with land and longstanding notions of trust and responsibility towards those dependent on the farm, the peasants have been stripped of their social world by state intervention into traditional institutions of social exchange. New formal organizations introduced by the state, that is, panchayats and above them the highly patronized and powerful factional alliances which have emerged from democratization process, have swept aside the consensual environment of the 'village republic'. The flagrantly corrupt rulers have become new trendsetters for the new generation of peasants' sons and daughters who have seen the traditional authority of their fathers and grandfathers being taken away by the 'gun-trotting militia men', patronizing politicians with huge amounts of finance at their disposal. Why should they toil at farms while others are making quick bucks and exhibit it in the form of palatial mansions? How can they compete with them?

Another collective institutional role needs consideration—the Akal Takht—as a repository of trust and imagined sovereignty of the Sikhs. The Takht was not only damaged physically in 1984, its demolition and rebuilding opened a deep psychological chasm within the community's shared values. With extreme pressure on its personnel in the immediate aftermath of the 1984 events, its authority was undermined first by the state authorities and then by the contingencies of Akali leaders, the supposed guardians of the Takht's sanctity. For ordinary Sikhs, it meant a constant double-speak on the part of its leaders—the exemplars of community's ethical standards. In their struggle to retain or to remain in political power at the provincial level, the Akalis have made such compromises and gone back so often on their promises that even the short memory of the public cannot ignore its devastating impact on the Sikh institutions of trust and exchange. During the last decade, the Akal Takht through its *Jathedar* (head man) has been used as nothing more than a political ally by the dominant Akali Dal faction. This has meant that the individual Sikhs' will and commitment to honour the faith's traditions have been compromised. The argument goes, if an apparently guilty person or party can be excused by the Akal Takht custodians, why should ordinary Sikhs be expected to hold 'ideals' when faced with economic and social survival? A few lies about their personal details to immigration authorities are justified in exchange for permanent settlement abroad, even if it means compromising social traditions, subverting social obligations of marriage governing the bride-givers and takers. The argument and psychological explanation that seems to run through, though never made explicit, is that if leading Sikh statesmen can speak one language in Delhi and another among their followers without impunity, why should ordinary persons feel such compunction in compromising their ethical standards? Thus, sacrificing conventional marriage customs are justified in order to settle abroad, escape the larger evil of 'bad governance' and accept marriage as a commercial transaction in order to export sons and daughters abroad.

Thus, a well-respected Sikh family that would have never contemplated marrying its son or daughter with a maternal cousin before the 1980s, finds itself ready for such a proposal solely to gain immigration. And payment for a bride or a bridegroom has become acceptable as part of changed social values. Post-colonial regime marked by ill governance has therefore generated a crisis for the Sikh peasantry leading to severe changes in its traditional norms of social exchange. The modernization experience has contributed to the Sikhs' exodus from Punjab in the form of emigration. The characteristic reaction of Punjabi peasants to the state's misgovernance may be interpreted by what Scott identifies as 'weapons of the weak' against 'high modernism' imposed from above. Why the Sikh peasantry's resistance has taken this particular form of flight from the province, than say another kind expression, is something beyond the scope of this paper and more question fit for psychological analysis (Scott, 1976; 1985; 1999).

CONCLUSION

While examining the changed nature of Sikh migration in the last three decades, this paper proposes an alternative hypothesis. It is suggested that the migration process must be linked to the modernization experience of the Sikh peasantry, and in particular due to their perception of being stifled by Punjab's political and social system. In seeking an explanation for rural Sikhs' reasons for leaving Punjab, the role of governance emerges as a crucial factor. That the state has failed to deliver 'good governance' to a highly politicized and educated peasantry is one of the chief reasons for the Sikh peasantry's collective behaviour appearing as mass migration. Nepotism, and a highly obstructive bureaucracy have turned Sikh peasantry's optimism and hopes of the Green Revolution into 'despair'. In the post-1984 period when death stalked the rural countryside the state effectively turned into a 'rogue state' adversely affecting the lives of thousand of young Sikhs and their hopes for the future.

This paper suggests that Punjabi migration in recent years ought to be studied in a wider context of the socio-political framework of the post-1966 Punjab. An explanation found in the Sikh peasantry's socio-psychological experience of modernization and their sense of governance experience in the Punjab should provide a more satisfactory analysis. Only by integrating collective behaviour and perception of Sikh peasantry can we construct a better theory for the last three decades of emigration process from Punjab. With a larger sample of oral histories of people settled abroad, as well as of those in the process of migrating, this is an endeavour for further research.

NOTES

1. *Hindustan Times*, 26 December 2004, pp. 1–2. The Malta tragedy occurred on the night of 25 December 1996 when 289 illegal migrants from India, Pakistan and Bangladesh were reported to have drowned. Among them a majority were Sikhs from Doaba. Of the 22 Punjabis who survived, 18 reportedly have tried again to emigrate. Balwant Singh Khaira is coordinating a legal battle for compensation for Malta tragedy victims in various European courts.
2. *Tribune*, 2 February 2005. Earlier, Captain Amarinder Singh, chief minister of Punjab, brought home 60 such stranded youth jailed in Pakistan as part of a goodwill mission for cordial relations between the two Punjabs—Indian and Pakistan.
3. Jalandhar city in the heart of Doaba region has over two hundred agencies just around the main bus stand with a further network of active canvassers and agents in the rural areas. A sample of the headlines appearing in the newspapers are: *The Tribune*, 7 March 2005: 'Master Cheats Caught in Police Dragnet'; 12 March 2005: 'Girl Marries Despite Knowing NRI Groom's HIV Positive' (this is a case of an Australian Sikh from Patiala marrying a young girl); 13 March 2005: 'Marriage Bureau Owner, 2 Others Booked for Cheating'; 2 March 2005: 'He Lost Mother, Money to "Land of Opportunities" '; 23 March 2005: 'Travel Agency Vanishes with Money, Passports'; 26 March 2005: 'Punjabi Youths Top Deportees' List'; 29 August 2005: '57 Youths Cheated by Travel Agents; Gang Busted'. *The Tribune*, 3 April 2006 reported 120 youth duped by travel agents who promised them jobs as drivers in Kenya. The agents in this case were reported to have vanished with Rs. 2 crore of would-be-migrants' money.
4. One can ask rhetorically; what is common between the following: Kapur Singh, the Sikh state protagonist; Tejwant Gill, a literary critic; Sucha Singh

Gill, an economist; and several of Punjab's royal families? The answer is they all have sent their off-springs abroad. Not for education or for a sort of sight-seeing but for permanent settlement. Here is the second puzzle. What is common between Narinder Singh Kapany, Jagjit Singh Chohan, the pop-singer Malkit Singh, creative writer Amarjit Chandan, Kesar Singh, a chronicler of the Ghadar movement through episodic fiction, Ujjal Dosanjh, a minister in the Canadian Federal government and ex-premier of British Columbia? They all are settled abroad but still care for Punjab profoundly, though differing in approaches about how this can best be done. The issue, of course, arises: what has forced so many Sikhs to emigrate from the land for which they continue to nurture so much affection?

5. Reports of such marriages make frequent news in the Punjabi press. See *Ajit*, 3 February 2005. During such a 'packaged' marriage in Jagraon, district Ludhiana, the 'bride' was arrested along with her 'fictitious parents' and other 'guests'. It was revealed that it was her fourth marriage posing as a Canadian girl, and the whole package cost Rs. 20 lakh for the family of the bridegroom who were a 'respectable Sikh family'.
6. For an exception see, Joyce Chaudhri (see Pettigrew) (1971), *The Emigration of Sikh Jats from the Punjab to England*, London: Final report on SSRC Project 331/1. Also, Archana B. Verma (2002), *The Making of Little Punjab in Canada: Patterns of Immigration*, New Delhi: Sage. Verma takes account of caste and *biradari* (Mahton) implications of migration from a village Paldi in Hoshiarpur.
7. The culture of migration is defined as 'a set of interrelated perceptions, attitudinal orientations, socialization processes and social structure, including transnational social networks, growing out of the international migratory experience, which constantly encourage, validate and facilitate participation in this movement'. See Wayne Cornelius (1992), 'From Sojourners to Settlers: The Changing Profile of Mexican Immigrants to the United States', in Jorge Bustamante et al. (eds.) (1992), *US-Mexico Relations: Labor Market Interdependence*, Stanford: Stanford University Press.
8. *Des Hoya Pardes* is a film made by Gurdas Mann, Punjab's popular singer.
9. *The Tribune*, 3 April 2006, reported 2,116 suicides between 1998 and 2004 while according to Jaijee, in one block of Sangrur district alone, there were 150 suicides, citing the total for Punjab to be as high as 40,000. The paper reported that Punjab's chief minister, Amarinder Singh, was seeking federal government assistance in solving the debt crisis of the peasantry. *The Tribune*, 16 April 2006, reported that 36 per cent of peasants want to give up farming and that suicides rates are likely to increase in case of governmental indifference to peasants' plight.

10. A preliminary survey of Sikh diasporan's philanthropy in Punjab shows a wide range of religious, social, and cultural institutions being supported by overseas Sikhs. See Verne Dusenbery and Darshan S. Tatla.
11. From 1983 onwards Punjab was placed under direct rule from Delhi, an Akali government led by Surjit Singh Barnala was dismissed after a year, then a Congress government led by Beant Singh was constituted with elections which were boycotted by Akali Dal in 1992. Compare the legitimacy of post-colonial state elected by 'people' with the colonial state imposed by 'foreigners'. More precisely, the contrast with a colonial official could not have been more striking: an English administrator on tour on a horse and attendants who would leisurely put up a tent in the evening near a village and ask its elders to gather to apprise him of their complaints and grievances and generally gauge the local opinion of the governed. See Malcolm Darling (1977), *Punjab Peasant in Prosperity and Debt*, Delhi: Manohar, reprint of 1925 edn.
12. The World Bank report cites the ratio of civil servants per 100 of Punjab population at 2.2 with all India ratio of 1.4. Top civil servants in absolute numbers grew from 612 in 1967 to 5,192 in 1992 and 12,156 in 2002, leading to a ballooning wage bill. For every single IAS officer in Punjab there are 412 employees of Group D staff (drivers, peons, etc.), compared to 352 in Tamil Nadu and 205 in Karnataka.
13. At the top echelon, the number of IAS officers swelled from 165 to 206. Between 1984 and 1998 Superintendents of Police and ranks above these increased from 112 to 237. Punjab had only one Inspector General (IG) of Police two decades ago, currently it employs 16 IGs, 13 Additional Deputy IGs with a Director General of Police at the top.
14. *The Tribune*, 1 January 2005 reported that in 2004, 66 gazetted officers and 165 non-gazetted officers were booked by the Punjab Vigilance Bureau on the allegations of being involved in corruption. Some 122 criminal cases were registered during the year against these officers. Earlier in 2003, the chief of Punjab Public Service Commission was found to have accumulated more than Rs. 100 crore through bribes in appointment to government jobs.
15. On an average day 36 per cent of primary teachers are absent from schools and of those in school, only half are teaching. In the health sector nearly 39 per cent doctors and 44 per cent other medical staff are absent from their place of work on any day. Punjab spends only 6 per cent of its budget on education and health. The number of surplus employees in the irrigation department is estimated at 20,000. The Punjab State Electricity Board is the most bloated electricity utility employing 85,000, that is 16 employees per 1,000 customers, double the Indian average.

16. For instance, for the year 2003, over 40,000 qualified teachers in Punjab were seeking jobs, many unemployed for over five years. Paradoxically, over 1,000 schools at the primary and secondary levels were short of teachers.
17. In technical education, for example, there were just 12 colleges in 1980; in 2004 the number had gone up to 83.
18. Although many countries admitted them, only a small proportion claiming refugee status were accepted. See reports of various refugee boards, especially of Canada, the United States and the UK.
19. This was especially so in western Europe; Germany and Belgium owed more to this factor during the late 1980s and early 1990s.

REFERENCES

Barrier, N.G. (1964), *The Punjab Alienation of Land Bill of 1900*, Durham: North Carolina, Duke University.

Boyd, Monica (1989), 'Family and Personal Networks in International Migration: Recent Developments and New Agendas', *International Migration Review*, 23, pp. 638–70.

Cliffe, L. and R. Luckham (1999), 'Complex Political Emergencies and the State: Failure and the Fate of the State', *Third World Quarterly*, 20(1).

Gill, S.S. and R.S. Ghuman (2000), 'Crisis of Punjab Economy: The Alternative Options and Role of the Government', in R.S. Bawa and R.S. Raikhy (eds.), *Punjab Economy: Emerging Issues*, Amritsar: Guru Nanak Dev University, pp. 437–58.

Green, P. and T. Ward (2004), *State Crime: Government, Violence and Corruption*, London: Pluto Press.

Hetchter, M. (1987), *Principles of Group Solidarity*, Berkeley: University of California Press.

Iyer, K.G. and M.S. Manick (2000), *Impoverishment and Suicides in Rural Punjab*, Delhi: Indian Publishers.

Jodhka S.S. (ed.) (2001), *Community and Identities: Discourse on Culture and Politics in India*, New Delhi: Sage.

——— (1997), 'Crisis of the 1980s and Changing Agenda of "Punjab Studies": A Survey of Some Recent Research', *Economic and Political Weekly*, XXXII (6), pp. 273–9.

Kumar, Promod and S.L. Sharma (1998), *Suicides in Rural Punjab*, Chandigarh: Institute for Development and Communication.

Mukherjee, Bharati and Clark Blaise (1988), *The Sorrow and the Terror: The Haunting Legacy of the Air India Tragedy*, Canada: Penguin Books.

Pettigrew, Joyce (1995), *The Sikhs of Punjab: Unheard Voices of State and Guerrilla Violence*, London: Zed Books.

——— (1971), *Report*, pp. 40–55.

Scott, James C. (1999), *Seeing Like a State: How Certain Schemes to Improve the Human Condition have Failed*, Yale: Yale University Press.

——— (1985), *Weapons of the Weak*, Yale: Yale University Press.

——— (1976), *The Moral Economy of the Peasant: Rebellions and Resistance in Southeast Asia*, Yale: Yale University Press.

Shergill, H.S. (1998), *Rural Credit and Indebtedness in Punjab*, Chandigarh: Institute for Development and Communication.

Shiva, Vandana (1991), *The Violence of the Green Revolution: Ecological Degradation and Political Violence in Punjab*, London: Zed Books.

Singh, Birinder Pal (2002), *Violence as Political Discourse: Sikh Militancy Confronts the Indian State*, Shimla: Indian Institute of Advanced Study.

Snowdon, L.L. (1990), 'Collective Versus Mass Behaviour: A Conceptual Framework for Temporary and Permanent Migration in Western Europe and the United States', *International Migration Review*, 24 (3), pp. 577–90.

Tatla, Darshan S. (2004), 'Writing Prejudice: The Image of Sikhs in Bharati Mukherjee's Writings', in Pashaura Singh and N.G. Barrier (eds.), *Sikhism and History*, New Delhi: Oxford University Press.

Turner, R.H. and L.M. Killian (1957), *Collective Behaviour*, Englewood Cliff: Prentice-Hall.

Voigt-Graf, Carmen (2003), 'Indian at Home in the Antipodes: Migrating with Ph.Ds, Bytes or *Kava* in Their Bags' in Bhikhu Parekh et al. (eds.), *Culture and Economy in the Indian Diaspora*, London: Routledge, pp. 142–64.

World Bank (2004), *Resuming Punjab's Prosperity: The Opportunities and Challenges Ahead*, Washington: The World Bank.

CHAPTER 14

Johl Diversification Scheme: A Critique

Harjinder Singh Shergill

Traditionally Punjab had been an agricultural economy. After the Partition of the country the most fertile part of the land went to Pakistan and what remained with India was least productive. The then eastern Punjab constitutes the present states of Punjab, Haryana and Himachal Pradesh.

India was a food deficit country in the early 1950s. On an average 2.5 million tonnes of food grains were imported every year till the middle of the 1960s. In 1964–5 alone the quantum of import increased to more than 13 million tonnes. This was a heavy loss of foreign exchange for the country. To enhance agricultural production under the PL-480 scheme, Punjab Agricultural University was established in Ludhiana to introduce modern scientific technology in agriculture, thus initiating the phase of Green Revolution in Punjab and elsewhere in India.

As a result of this revolution, per capita availability of cereals in the country increased from 334.2 gm per day in 1951 to 417.6 gm in 1971, 417.3 gm in 1981, 468.5 gm in 1991 and finally declined to 390.6 gm in 2001. Punjab with only 1.53 per cent of the geographical area of the country was producing only 3 million tonnes of food grains in 1961 that increased to 7.3 million tonnes in 1971, 19.2 million tonnes in 1991 and to 25.3 million tonnes in 2001. Thus, from a food deficit country, India came to acquire huge stocks of 70 million tonnes in 2002 when the requirement was only for 22 million tonnes.

The diversification of agriculture in Punjab and the question of shifting area out of wheat and rice cultivation to other crops has been hanging in fire for the last two decades. In the first diversification report (Johl, 1986) commissioned by the Punjab government, the expert committee had strongly recommended the replacement of wheat and rice by other crops on a sizeable cultivated area of the state. However, the committee did not suggest any concrete scheme for inducing farmers to shift land out of wheat and rice cultivation. So no follow-up action was taken by the government. Since then (1986), Punjab farmers have further expanded area under rice by 10 lakh hectares and under wheat by about 3 lakh hectares. The actual evolution of cropping pattern in the state has gone in the opposite direction to what was recommended by the expert committee.

In the second diversification report (Johl, 2002), the expert committee gave a detailed scheme for shifting 10 lakh hectares of cultivated area out of wheat and rice cultivation to alternative crops. The scheme called Crop Adjustment Programme (CAP) has been accepted and adopted by the Punjab government as the main plank of its agricultural policy. The scheme has been submitted to the central government for approval and funding. If approved and implemented, it will have far-reaching and multifarious effects on Punjab agriculture. In view of the big size of the shift in area proposed, and the official adoption of this scheme by the Punjab government, the CAP is critically examined in the next few pages.

A SYNOPTIC VIEW OF THE JOHL DIVERSIFICATION SCHEME

For a critical assessment, it will be useful to start with a brief description of the CAP proposed by the Johl Committee. The description is also necessary in view of the difficult availability of the Johl Committee report to the lay public, and even to most academicians. Most people have gathered bits of information about the scheme from what has been appearing in newspapers and only a few have actually read it.

Objectives of the Scheme

The main objectives of this scheme, as outlined in the Johl Committee report, are the following:

1. To shift 10 lakh hectares of cultivated area of the state from under wheat and rice to other crops, preferably oilseeds and pulses. This shift in area will reduce production of wheat by 46.9 lakh tonnes and that of rice by 33.0 lakh tonnes.
2. To save the funds and effort that the central government spends on procurement, handling, transport and storage of about 80 lakh tonnes of wheat and rice produced and marketed from these 10 lakh hectares of cultivated area of Punjab.
3. To check the decline in water table in the state by reducing the withdrawal of ground water used for irrigating wheat and paddy on these 10 lakh hectares of cultivated area.
4. To save the agro-ecological environment of the state by reducing the soil, water and air pollution caused by excessive wheat and rice cultivation.
5. To ease congestion in the grain markets of the state by reducing market arrivals of wheat and paddy by about 97 lakh tonnes.

Mechanics and Implementation of the Scheme

Under CAP the farmers who agree to shift cultivated area from wheat and paddy will be paid Rs. 12,500 per hectare per year as compensation for the deficiency in their incomes caused by this shifting of area to other crops. The village panchayats will be paid Rs. 250 per hectare for working out the implementation of this programme in their respective villages. The state government will be paid Rs. 5 crore as administrative, management and supervision costs of the scheme. The total amount under the three heads comes to Rs. 1,280 crore per year and is expected to be provided by the central government for the implementation of this scheme.

At the ground level the scheme will be operated and implemented jointly by the village panchayat, village *nambardar* and *patwari* on the basis of information on area under various crops given in the *khasra girdawari* records. First the area at present under wheat and paddy will be marked out. Then the land that a farmer agrees to shift from under wheat and paddy will be marked. Preference will be given to farmers who agree to shift their entire landholding from under paddy and wheat to other crops. The farmers participating in the scheme will be paid half the compensation before the sowing of wheat and paddy, and the other half of the compensation will be paid at the end of the crop season after making sure that the participating farmers have actually shifted the marked area. There will be deterrent penalty for violation of the agreement by the farmer. In case of violation of contract by a farmer, double the amount of compensation will be recoverable as arrears of land revenue. At the state government level, the scheme will be implemented and supervised by the Directorate of Agriculture through its district-level and block-level agricultural development officers.

Benefits of the Scheme

The CAP for diversification of Punjab agriculture is supposed to benefit all the parties: the central government, the state government and the farmers. The benefits to these three parties, as mentioned in the Johl Committee Report (2002) are detailed below.

1. *Benefits to the Central Government*

The implementation of this scheme will result in a net financial saving of about Rs. 5,000 crore to the central government. At present, that is, the year 2002 when the scheme was proposed, the central government is spending about Rs. 9,000 crore per year on the procurement, handling, transport and storage of 80 lakh tonnes of wheat and rice produced on the 10 lakh hectares of area proposed to be shifted out of wheat and paddy cultivation. Even after paying Rs. 1,280 crore to the Punjab farmers for shifting 10 lakh hectares of area from under wheat and paddy cultivation, the central government can save about Rs. 5,000 crore annually. Further, the oilseeds and pulses produced on this 10 lakh hectares of area will drastically reduce the imports of oils and pulses, and thus save huge amount of foreign exchange.

2. *Benefits to the Punjab Government and the State*

The pressure on the state government and its agencies for the procurement of wheat and paddy will be reduced by the reduction in wheat and paddy output in the state by about 97 lakh tonnes. The handling, transport, credit and storage arrangements for the remaining smaller output of wheat and paddy will be easier for the state government. The government will be relieved of the pressure from farmers' unions for higher Minimum Support Price (MSP) every season. The fall in water table in the state will also be checked. Soil, water and air pollution will be reduced to tolerable levels and the ecological environment of the state will improve.

3. *Benefits to the Punjab Farmers*

The participating farmers will get Rs. 12,500 per hectare as income support for making investments on their farms. The irrigation, power and labour requirements for farming will be reduced and ease the life of farmers from chronic shortages of these inputs, particularly in the peak sowing season. The problems faced by farmers due to glut in grain markets will also be eliminated. The use of machinery on the farm will be normalized over the years and thus farm efficiency will be improved.

4. *Other Benefits: WTO Compatibility and Flexibility*

Another benefit of this scheme is that the deficiency payments made to farmers under this scheme are compatible with the WTO stipulations, because this support is not trade distorting subsidy and falls in the Green Box category. Further, the scheme is flexible and enables the government to increase or decrease area under wheat and paddy according to the needs of the nation for food grains. The area adjustments can be triggered almost instantly; it can be put on and off as per the needs of the market.

Necessary Conditions for Implementation

For successful implementation of the proposed CAP, the following necessary conditions are mentioned and emphasized in the Johl Committee report:

1. The central government will have to provide a subsidy of Rs. 1,280 crore per year to Punjab for shifting 10 lakh hectares of area from under wheat and paddy cultivation.
2. The Minimum Support Prices and Assured Purchase (MSPAP) programme for wheat and paddy will continue without any change.
3. An effective MSPAP programme will be introduced for the alternative crops (i.e. pulses and oilseeds) that replace wheat and paddy on the shifted 10 lakh hectares of cultivated area.
4. An adequate marketing and processing infrastructure for the crops that replace wheat and paddy has to be created fast in order to sustain the cultivation of these crops.

This is the broad description of the bold scheme for diversification of Punjab agriculture prepared by the Johl Committee (2002) adopted by the state government and submitted to the central government for approval and funding. To date, it has neither been approved nor completely rejected by the central government. It is still under consideration.

EVALUATION OF THE JOHL DIVERSIFICATION SCHEME

The CAP proposed by the Johl Committee is based on the inspiration and ideas drawn from the well-known 'land retirement programmes' that have been in operation in developed countries like the USA, the UK and the European Union for almost half a century now. The long and varied experience of these countries in operating the 'land retirement programmes' has generated enough empirical evidence on their strengths and weaknesses in different agro-climatic and socio-economic regions. The lessons learned from the experience of developed countries can be helpful in evaluating the benefits, costs and workability of the diversification scheme proposed by the Johl Committee. The Johl diversification scheme is, in fact, a poor parody of the standard 'land retirement programmes' in operation in the rich countries. The scheme seems to have been drawn up in a hurry, without bothering about its workability and various ramifications. The enthusiasm for quick solutions and for making it attractive to the central government has prevented its authors from weighing its benefits, costs and workability in the current socio-economic and political environment of the country and the state. In the next few pages we discuss the obvious, direct and indirect effects of this scheme to replace wheat and paddy by other crops on 10 lakh hectares of Punjab lands. The feasibility and workability of this scheme under the present parameters and constraints is also examined.

I. Income Loss from Shifting Area out of Wheat and Paddy Cultivation

The crucial factor in shifting area out of wheat and paddy cultivation is its impact on the incomes of farmers and other classes engaged in and dependent on agriculture. But the scheme is silent about the likely impact on farmers' income if 10 lakh hectares are shifted out of wheat and paddy cultivation; only the compensation of Rs. 12,500 per hectare to be paid to farmers is mentioned in the scheme. The likely negative impact on farm incomes and on incomes of other rural/urban strata dependent on agriculture are presented in the next few paragraphs.

1. *Fall in Farmers' Income*

Our estimates indicate that farmers will suffer a net loss of Rs. 13,280 per hectare per year by shifting out of wheat and paddy cultivation, even after receiving the proposed cash compensation (subsidy) of Rs. 12,500 per hectare. This estimate is based on the data given in the 'CAP' scheme itself (Johl, 2002, Table 1.3, p. 20) and pertains to the year 2001–2. The details of the estimate of income loss to farmers are given in Table 14.1. From these estimates it is clear that Punjab's farmers will suffer a net loss of Rs. 13,280 for each acre shifted out of wheat and paddy cultivation and the total income loss to farmers as a class comes to Rs. 1,328 crore, if 10 lakh hectares are shifted out of these two crops. Given such a big loss in the farmers' income, it is certain that willingly farmers will not shift any area out of these two crops, notwith-

TABLE 14.1: FARMERS' NET LOSS PER HECTARE SHIFTED OUT OF WHEAT AND RICE CULTIVATION (REFERENCE YEAR 2001–2)

Item description	Estimated amount (Rs. in crore)
1. Loss in income from earnings foregone from one hectare shifted out of wheat and rice	31,222
2. Compensation (subsidy) paid to farmers under the scheme for shifting one hectare out of wheat and rice	12,500
3. Net earnings from alternative crops (pulses and oilseeds) grown on one hectare shifted out of wheat and rice	5,442
4. Net loss to farmers from one hectare shifted out of wheat and rice 4 = (1) – (2 + 3)	13,280
5. Net loss from 10 lakh hectares shifted out of wheat and rice (5) = (Rs. 13,280) × 10 lakh hectares	1,328

Source: Johl Committee Report, 2002.

standing the offer of a cash compensation of Rs. 12,500 per hectare under the scheme.

2. *Fall in Earnings of the Farm Labour*

The main source of employment and earnings of agricultural labourers in Punjab is wheat and rice cultivation. Considerable amount of hired labour is used in the cultivation of these crops. The shifting of 10 lakh hectares of area from under wheat and paddy will result in a big fall in the demand for hired labour and consequently in the earnings of farm labourers. Our estimates for the year 2001–2 indicate that agricultural labour class in the state will suffer a net loss of Rs. 543.23 crore in their annual earnings, if 10 lakh hectares are shifted out of wheat and paddy cultivation. These estimates are based on the amount of hired labour used and wages paid in wheat and paddy cultivation and other crops, as given in the 'Cost of Cultivation Reports' published by the Commission for Agricultural Costs and Prices (CACP), Government of India. The details of these estimates are given in Table 14.2. These

TABLE 14.2: NET EARNINGS LOSS TO THE FARM LABOUR (REFERENCE YEAR 2001–2)

Item description	Amount (Rs. in crore)
1. Earnings loss to labour from 10 lakh hectares shifted out of wheat	367.79
2. Earnings loss to labour from 10 lakh hectares shifted out of rice	327.70
3. Earnings gain to labour from 10 lakh hectares put under alternative crops (pulses and oilseeds)	152.26
4. Net earnings loss from 10 lakh hectares shifted out of wheat and paddy (4) = (1+2) – 3	543.23

Source: Based on information on the use of hired labour in different crops given in the CACP reports.

estimates indicate that the class of farm labourers in the state will suffer a net income loss of Rs. 543.23 crore per year if 10 lakh hectares of area is withdrawn from wheat and paddy cultivation. Given that a large number of farm labour families are only marginally above the poverty line, such a big loss in their earnings will push many of them below the poverty line.

3. *Fall in Earnings of Other Occupational Groups*

Farmers and farm labourers will not be alone in suffering a major income loss from this move to divert 10 lakh hectares from wheat and paddy cultivation. Other occupational groups involved in harvesting, marketing and handling of wheat and rice will also suffer substantial loss in their incomes. The income loss estimates for some of these social groups have been prepared on the basis of information available in the Johl Committee Report (2002) and in other secondary sources. These estimates are presented in Table 14.3 and indicate that these other occupational groups, involved in wheat and rice production in several ways, will suffer an annual income loss of Rs. 907.75 crore if 10 lakh hectares are shifted out of wheat and rice cultivation.

TABLE 14.3: INCOME LOSS TO OTHER OCCUPATIONAL GROUPS (REFERENCE YEAR 2001–2)

Item description	Estimated amount (Rs. in crore)
1. Harvester combine and tractor owners hiring out machinery	446.99
2. Commission agents	149.52
3. Marketing committees and Mandi Board	119.61
4. Labour working in the grain market	73.41
5. Transport/truck owners involved in grain transporting	118.22
6. Total loss to all these five groups (6) = (1+2+3+4+5)	907.75

Source: Based on information given in the Johl Committee Report, 2002.

4. *Total Income Loss to the State as a Whole*

The combined direct negative effect on income of farmers (Rs. 1,328 crore), farm labourers (Rs. 543.23 crore), and other occupational groups (Rs. 907.75 crore) comes to Rs. 2,778.99 crore per year, if 10 lakh hectares are shifted out of wheat and paddy cultivation. Putting this income loss in relative terms, it will result in 17.54 per cent fall in net state domestic product (SDP) originating in the agricultural sector. Such a big fall in income will plunge the state economy in economic recession through a chain reaction of multiplier effects. Depending upon the value of multipliers, the final income loss to the state will be much higher than Rs. 2,778.99 crore.

5. *Other Recessionary Effects*

In addition to the direct income loss of Rs. 2,778.99 per year mentioned above, the reduction of 10 lakh hectares in area under wheat and paddy will also have a number of other indirect income depressing effects. For example, the sale of fertilizers in the state will fall by Rs. 626.39 crore per year, and that of other chemicals and insecticides by Rs. 131.54 crore. About one-third of the rice shelling capacity in the state will be rendered idle, with attendant loss of employment and earnings to labour working as rice shellers.

(Even the state government will suffer a financial loss from the reduced fees earned by it from public sector agencies for handling procurement and marketing operations on their behalf.) There is thus little doubt that the shifting of 10 lakh hectares from under wheat and paddy cultivation will have an all-pervasive and enormous recessionary impact on incomes in the state.

II. Unrealistic Time Frame and Size of Area Reduction

The sheer size of reduction in area (10 lakh hectares) under wheat and paddy proposed in the Johl diversification scheme is mind-boggling. The small time frame (one year) for this shift in area makes the scheme simply impossible. In their overenthusiasm for quick diversification of Punjab agriculture, the authors of the scheme did not see the gigantic size of the shift in area they were proposing. In the year 2001–2, the total area under wheat in Punjab was 42.6 lakh hectares and under paddy 26.1 lakh hectares. The reduction in area under wheat and paddy by 10 lakh hectares in one year means a 23 per cent reduction in the present area under wheat and a 38 per cent reduction in area under paddy. The enormity of this proposed shift in area becomes clear when we see it in the historical context of expansion of area under these two crops in Punjab. It means reverting the area under paddy in one year to what it was in the year 1984–5; and reverting the area under wheat back to where it stood in the year 1975–6. The expansion in area under these two crops that occurred gradually over decades is sought to be reversed in just one year; only a magical trick can accomplish such a great feat. The available literature on this subject suggests that such a huge reduction in area under any crop, in such a short time (one year) has never been attempted anywhere, not even in the developed countries which invented the 'land retirement programmes'. A quantum leap backwards of this huge dimension seems impossible to achieve in one year; even if both the governments—state as well as the central—and the farmers are keen and enthusiastic to make such an attempt. The Johl diversification scheme is doomed by its gigantic size and small time frame. But given the big income loss to farmers by shifting area from under wheat and paddy, they cannot be a willing party to their own economic ruin. Rather than reducing area under wheat and paddy, the Punjab farmers are expanding area under these two crops every year. Thus, it is clear that the enormous size of reduction in the proposed area and too small a time span in which it is expected to be achieved makes this scheme totally unrealistic, with little chance of success even if the central government agrees to provide funds for its implementation. But no central government will be willing to fund reduction in area under wheat and rice in view of its serious implications for the country's food security.

III. Incompatibility with Country's Food Security

Given the political reality of Punjab being a part of the Indian Union, no agricultural policy for the state has any chance of succeeding if it is not compatible with the food security of the country as a whole. All reasonable estimates of food demand and supply balance in the country suggest that India will be barely sufficient in food grains production up to the year 2050. On account of growth of population and per capita incomes, demand for food grains will grow at a rate with which growth of food grains production will find difficult to keep pace with. Most of the growth factors such as expansion in cropping area, irrigated area and growth of crop yields are already exhausted. Some studies even predict that India will be a net importer of food grains by 2050. Given such a dismal future of food grains availability, no central government is going to fund a scheme to reduce food grains production in the country. The Johl diversification scheme is not attractive to the Punjab farmers even with a central government cash subsidy of Rs. 12,500 per hectare; but without such a central government-funded subsidy there will be simply no takers for this scheme.

IV. Preconditions of the Scheme Cannot be Fulfilled

For successful implementation of CAP, a number of preconditions are laid down by the authors of the Johl Committee Report. The more important of these necessary conditions have been described in the previous section. Unfortunately, none of these conditions can be fulfilled in the present political and economic scenario of the country. The central government is not agreeable to a drastic reduction in food grains production in the country, not to talk about funding such a move. Further, the central government cannot agree to start an entirely new deficiency payments programme for wheat and paddy due to many reasons. Once started in Punjab, it will become a precedent for other states which will also demand similar programmes for their farmers. For instance, Maharashtra will ask for reducing area under sugar cane. Moreover, in the current situation of fiscal crisis the central government is simply not in a position to provide funds needed to introduce this scheme.

Similarly, the central government is not willing to start any new MSPAP programme for the alternative crops being recommended to replace wheat and paddy. It is rather searching ways, means and excuses to wriggle out of the existing MSP programme for wheat and paddy.

The processing, marketing and storage facilities in the state for the alternative crops that will replace wheat and paddy are quite inadequate and cannot be quickly created in the public sector because of the continuing financial crunch. There is little hope that private sector investors will create these processing and marketing facilities quickly as production of these alternatives crops expands under this scheme. It is thus clear that none of the preconditions for the success of this scheme, as laid down by its authors, can be fulfilled in the near future. In view of that the prospects of the scheme embracing success are indeed bleak.

V. Administrative Burden of the Scheme Makes it Impossible to Implement

The administrative demands of this scheme are staggering. Instead of reducing the role of government in the management of marketing of farm produce, it will increase the workload of public sector agencies and government departments considerably. Even after the implementation of the scheme, the existing MSPAP programme for wheat and rice will continue, though at a reduced level. A new comprehensive (covering pulses and oilseeds) MSPAP programme will have to be introduced and operated by the government agencies. An entirely new programme of direct cash payment totalling Rs. 1,280 crore per year has to be created and operated by the government to distribute this amount to lakhs of farmers spread over 12,000 villages of the state. So the involvement of government and its agencies in marketing management will not only increase considerably, but will also become more complex and hence more difficult. It is doubtful if the already overstretched and inefficient government apparatus will be able to cope with the additional administrative tasks necessary to successfully implement this scheme.

The implementation procedure outlined by the Johl Committee also creates serious doubts about the workability of this scheme. The involvement of about 10,000 *patwaris* and more than one lakh panchayat members and village *nambardars* (envisaged in the scheme) in the distribution of direct cash payment of Rs. 1,280 crore per year is a sure road to disaster. Given the state of land records, method of working of the *patwaris*, and inefficiency and factionalism of panchayats, one can imagine how much of the Rs. 1,280 crore will reach the actual farmers participating in the scheme; and how much of the diversified area shown in *girdawaris* will be actually retired from wheat and paddy cultivation. It is more likely that even after implementation of the scheme (on paper), the actual fall in wheat and paddy production may be only marginal. It is thus clear that given the complex and staggering administrative requirements and problems in implementing the scheme, its chances of success are indeed bleak.

VI. Area Reduction Programmes: Lessons from the Experience of Developed Countries

The area reduction schemes called 'land retirement programmes' have been in operation in many developed countries for almost half a century. The experience of running these programmes, even in the more favourable conditions of the developed countries with plenty of funds, has shown that these fail to attain the goals with which these are introduced. Rather, many negative side effects and externalities get generated. The reduction in production (the main goal) of a crop is far less than what is planned and paid for by the government under such a programme. Firstly, farmers retire/divert mostly marginal/low productivity lands from under the designated crop. Secondly, farmers start using the variable inputs (fertilizers, etc.) much more intensively on the remaining area under the crop because of softening of the cash constraint by the money paid to them under the programme. Thirdly, even the fixed inputs—machinery, tube wells, family labour and managerial skills—are used more intensively on the remaining area under the crop because the opportunity cost of using these fixed inputs is very low. All these economic mechanisms come into play as soon as an area reduction programme is introduced and as a result the reduction in output is much less than the production in which area under the crop is retired. In addition to the above purely economic mechanisms, outright cheating by farmers in connivance with government officials implementing the scheme makes area reduction more a paper exercise and not a ground reality. Farmers invent ingenious ways and subterfuges to get the cash compensation payments without reducing area under the crop by the agreed amount. The experience of developed countries of operating these 'land retirement programmes' has been frustrating and these are now being eased out even in the rich countries.

If this is the efficiency and performance of area reduction programmes in the much more favourable conditions of the rich countries with much lower levels of corruption, one can very well imagine the chances of their success in the less favourable environment of underdeveloped, poorly governed countries like India where corruption is rampant. In such an environment all the above mentioned ill effects and externalities will emerge in a much stronger form and make the proposed area reduction programme completely ineffective. A part of the cash compensation money will not even reach the farmers, and what reaches them will be pocketed by farmers without reducing area under wheat and paddy by the agreed amount. It is thus clear that these leakages will erode the impact of the proposed area reduction programme considerably.

CONCLUSION

From the foregoing discussion it is clear that farmers, farm labourers and others dependent on agriculture will suffer a big income loss if CAP is implemented. The central government cannot agree to its implementation and funding in view of its incompatibility with food security of the country. Besides, the scheme's administrative requirements are too heavy and complex and will make its implementation almost impossible. Even if implemented, large-scale leakages and cheatings will make it completely ineffective in reducing wheat and rice production by the amount envisaged in the scheme.

REFERENCES

Johl, S.S. (2002), *Agricultural Production Pattern Adjustment Programme in Punjab for Productivity and Growth*, Chandigarh: Government of Punjab.

——— (1986), *Diversification of Agriculture in Punjab*, Report of the Expert Committee, Chandigarh: Government of Punjab.

CHAPTER 15

Beyond 'Crisis': Rethinking Contemporary Punjab Agriculture

Surinder Singh Jodhka

Captain Kanwaljit Singh, the then finance minister while presenting the annual budget to the State Assembly of Punjab on 19 March 2001, underlined the need for addressing the question of Punjab farmers committing suicide. 'Such incidents of suicide', he argued, 'need to be identified and the families becoming destitute need support for their social and economic rehabilitation.' Making it a part of the state's budgetary provisions, he also announced that if the earning member of a family committed suicide, 'the state would identify the case and pay a compensation of Rs. 2.50 lakh for the rehabilitation of the family'.[1] Since a lot had already been written and talked about the impending agrarian distress/crisis, and a large number of suicides by marginal and small farmers had been reported from different parts of India, it was perhaps for the first time that a state government officially acknowledged the existence of such a phenomenon, and made some provision for helping the aggrieved families.[2]

Given that the Green Revolution was first introduced in Punjab, it also matured here before it did in other parts of the country. New technology raised land productivity by several times. It also changed the rural social structure and had a direct bearing on the nature of local and regional politics. Punjab emerged as a leading state of the country with highest per capita income.

However, the excitement did not last long. After two decades of growth, the Green Revolution began to lose its charm, and was followed by a series of crises. Beginning in the early 1980s, 'crisis' became the dominant mode of representing Punjab. From politics and economics to culture and ecology, everything seemed to be in a state of crisis in the state. The rise of Sikh militancy during the 1980s, for example, was often explained by referring to the successes and failures of the Green Revolution (Jodhka, 1997; 2001a).

While it might at times serve some useful political or even academic purpose to talk about the existing state of affairs in terms of crisis, such a language could also foil any recognition of the processes of social and economic change being experienced on ground, desirable or undesirable. It is precisely this that the present paper attempts to do, viz., look at the socio-economic changes that have come about in the rural Punjab and how best to explain the existing state of affairs in Punjab agriculture.

Located on the north-west border of India, Punjab is a rather small state, occupying less than 2 per cent of the total geographical area and inhabiting a little more than 2 per cent of the total population of the country. Bengal and Punjab were the only two provinces partitioned at the time of the formation of independent nation-states of India and Pakistan in 1947. Punjab thus became a border-state, located on the 'periphery' of India. The state was peripheral to India not only geographically but also socially and culturally. Ever since its reorganization in 1966, Punjab has been one of the few regions of the country where Hindus, constituting more than 80 per cent of India's population, have been a minority.

Notwithstanding its peripheral location, Punjab has always been an important region in the political and cultural imagination of the nation. As mentioned above, until recently Punjab was viewed as the most dynamic and progressive state of the country, mainly due to its successes in the agrarian sector. The Green Revolution was successful in other parts of India as well, but it was Punjab that it primarily came to be identified with.

The available statistics on various indicators of agricultural growth speak for themselves. Of all the states of India, Punjab's agricultural growth rate was the highest from the 1960s to the middle of the 1980s. Annual rate of increase in production of food grains during the period 1961–2 to 1985–6 for the state was more than double the figure for the country as a whole. The percentage of high yielding varieties (HYV) of seed in the total area under food grains in Punjab was as high as 73 per cent in 1974–5 (all India: 31 per cent) and 95 per cent in 1983–5 (all India: 54 per cent). While Punjab had 17,459 tractors per hundred thousand holdings, the all India figure was only 714. The same holds true for most other such indicators (Kohli and Singh, 1997). These achievements have been widely recognized. The opening lines of recent World Bank report on the state, for example, summarizes Punjab's achievements quite well:

> Punjab is India's most prosperous and developed state with the lowest poverty rate. At the end of the 1990s, more than 94 per cent of Punjab's citizens were above the poverty line, 70 per cent were literate, 94 per cent of the six year olds were enrolled in primary schools, 72 per cent of children under twelve months were immunized, 99 per cent of households had access to safe drinking water, and the average life expectancy of its citizen was 68 years. The remarkable development record of Punjab can also be inferred from the fact that it has already achieved, or is well on track to achieve, most of the Millennium Development Goals (MDGs). According to India's National Human Development Report (2001), Punjab was ranked second only to Kerala in terms of the overall level of human development among the major Indian states. Most citizens of Punjab have thus already achieved a level of socio-economic status that the majority of Indian citizens are unlikely to experience in their lifetime. (World Bank, 2004: 3)

Apart from the prosperity that the success of Green Revolution in the 1960s and 1970s brought to the people of Punjab, it also played a very important role in solving the gargantuan problem of food scarcity in the country. The state rightly came to be known as the food basket of India. The official website of the state government proudly claims that 'Punjab produces 22 per cent of the country's wheat (12.7 million tonnes), 9 per cent of rice (6.8 million tonnes) and 24 per cent of cotton (0.3 million tonnes). It contributes 60 to 70 per cent of wheat and 40 to 50 per cent of rice to the central pool.'[3]

The Green Revolution also changed the politico-cultural dynamics of the state. It was not only to the new agrarian technologies and the HYV seeds that the success of the Green Revolution was attributed. Credit was also given to the enterprising, rugged farmers of the region and their hard work. Their love for the land and the high values they attached to the practice of self-cultivation (*khudkasht*) played an important role in making the Green Revolution such a success in the region, much before it took roots in other parts of India.

The local dominant agrarian caste, the Jats (or Jutts) have particularly been known for the pride they take in their rural identities. A sociologist working in a village of the Doaba region during the early 1980s had reported that 'the Jat might be employed as a school teacher, or serve in military but he saw his primary role as that of an agriculturalist; his connection with land was what he held most dear and what identified him' (Kaur, 1986: 233). Another anthropologist similarly writes about the contempt that the Jats had for city life. They despised 'the townsman as lacking in physical bravery' and as 'grasping, greedy and lacking in dignity' (Pettigrew, 1992: 169).

As it happened in other parts of the country, the success of Green Revolution technology and the introduction of universal adult franchise brought the locally dominant castes to centrestage of the regional/state politics. This process was perhaps more intense in states like Punjab and Haryana where the agrarian lifestyle came to be the norm. The triumph of agrarianism not only gave the agrarian elite political power at the local and state levels but also had important implications for the existing social identities. For example, though the landowning Jats (Jutts) had always been an important element of the Sikh community, it was after the Green Revolution that the Sikh image came to be identified with the Jats (Jutts) (Gupta, 1996; Pettigrew, 1995). And, perhaps more importantly, despite its urbanization and industrialization being above the national average, Punjab came to be popular as a land of prosperous agriculturists.

THE 1980s AND AFTER

The decade of 1980s was a critical period in contemporary Indian history. Punjab witnessed a powerful ethnic movement during the 1980s. The movement for Khalistan, a separate Sikh nation, generated a sense of 'crisis', which was felt much beyond Punjab (Saberwal, 1987). Though Sikh militancy declined during the early 1990s, it had far-reaching consequences for the society and economy of Punjab.

Further, effects of the so-called 'Punjab crisis' were not confined to the region alone. The 'new social movements' that came up around the same time in different parts of the subcontinent, though very different in their content, also had several things in common with the crisis in Punjab. The 'new' mobilizations by women, farmers, dalits, tribals and ethnic communities, all questioned the wisdom of state-directed development and the 'Nehruvian agenda' of social transformation and modernization. Coupled with other changes at the global and national level, the decade of 1980s saw an overall erosion of the developmental state. As Das puts it:

> The goals of rational organization of life, the scientific management of society, modernization and development, to which great energies had been devoted in the sixties and early seventies, now seem like signposts to cities that are abandoned and empty. (Das, 1990: 1)

While the 1980s was, in a sense, a creative period for Indian society when many fundamental assumptions around which the post-colonial Indian nation was being built were questioned (Jodhka, 2001b), for Punjab and for the Sikhs it was a traumatic phase. Fifteen long years of militancy and the manner in which the Indian State handled the 'Punjab crisis' not only caused bloodshed and suffering, it also fundamentally altered the popular image of the region. From a region known for its economic vibrancy and progress, Punjab began to be seen as a 'crisis ridden state', a region with serious problems of law and order and political unrest and therefore not suitable for safe investments.

Interestingly, despite all problems, even during the 1980s the agrarian economy of Punjab continued to progress. The income of the primary sector of the state economy grew at an average of 5 per cent per annum while the corresponding figure for India as a whole was around 3 per cent.[4] The real implications of the crisis were to be felt in the following decade, during the 1990s, when the economic priorities at the national level witnessed a major shift.

GLOBALIZATION AND INDIAN AGRICULTURE

As discussed above, the decades 1980s and 1990s were important turning points in the history of contemporary India. It was perhaps for the first time in the post-Independence period that there was a general feeling of unease and doubt about the paradigm of development planning that the Indian State had embarked upon after its Independence from colonial rule. There had been criticisms of the policies and programmes that the first democratic government of independent India had initiated under the leadership of Pandit Jawaharlal Nehru earlier also, but they emanated mostly from conflicting ideological positions of different leaders or political formations. The strikingly new feature of the 1980s was that unlike before, the challenge this time came from 'below', from those who were supposed to benefit from development, or whom the independent Indian State had promised a better life.

In the following decade of the 1990s, the Indian State embarked upon a new framework of economic development. Pressed hard by the compulsion of changing global economy and rising import bills following the 'first' Gulf War, the Government of India initiated a process of economic reforms. Though initially intended to deal with the immediate challenge of 'balance of payment', the reforms turned out to be the beginning of a new phase in the economic history of India. The collapse of the Soviet Union and the end of Cold War around the same time revived the confidence of the advocates of free market economy. Breakthroughs in telecommunication technology and the increasing reach of capital to virtually every nook and corner of the world ushered in a different phase in globalization.

These changes also initiated certain new trends in social sciences. The 'old' modernist theoretical perspectives gave way to a variety of 'post-modernist' ways of imagining the world. Not that everyone in social sciences quickly converted to these 'new ways of looking at things', yet their presence was felt everywhere. The language of development discourses and politics of social change witnessed many shifts. From class analyses, the focus moved to questions of culture, from revolution to empowerment, from politics to governance, from production to consumption.

Though it may indeed appear paradoxical, but historically speaking, the rise of 'new social movements' in India and the trends that appeared in the following decade, liberalization/globalization, followed by the ascendance of a variety of post-modernist perspectives in the social science academy, all seemed to converge at some level with the idea of development that the new economic policies advocated. The connection between the rise of new paradigms in social sciences and the emergence of 'new' social movements was not entirely accidental. They did seem to support each other. They also seemed to agree on their criticisms of the modern state, and invariably projected it as a villain. They all seemed to also emphasize on a greater role for civil society institutions, which in effect meant opening up spaces in the sphere of development for non-governmental organizations (NGO). This kind of ideological shift of course suited the post-liberalization/globalization state well.

These shifts have had diverse implications for different categories of Indian population. For sections of the urban middle classes and the rich, the new economic policies have proved very rewarding. The NGO movement has also acquired a certain degree of respectability and strength. Questions of human rights, caste, gender, and ecology/environment have come to occupy the centrestage of social agenda in India and the world over, and can no longer be ignored by the state. However, these shifts have also marginalized certain 'old questions', questions that continue to be of critical significance, and have consequences for large masses of the Indian people.

The most obvious issue in this category is the marginalization of rural people in general and of those dependent on agriculture in particular. It is not only ideologically that agriculture has experienced marginalization in the popular imagination of the Indian people over the last two decades; its share in the national income has also declined considerably. Though a large majority of Indians continue to live in the countryside, the share of agriculture to the national income has come down to less than a quarter. The growth rates in agricultural sector have also been much slower than other sectors of the economy.

Declining significance of agriculture, one would think, is quite 'natural' and perhaps a desirable process. With development of industry and modern servicing sectors, it has happened everywhere in the world. However, there is something quite unique about the Indian experience. Unlike other regions of the world, marginalization of agriculture in the Indian economy is not being accompanied by a similar degree of shift of population to non-agricultural employment. Given that India is a democratic country such a reality becomes even more challenging.

LIBERALIZATION AND THE PUNJAB ECONOMY

Though liberalization and globalization were important turning points in the recent economic history of India, the 'crisis' of Punjab agriculture, as mentioned above, was already evident by early 1980s, and had become a political issue in the state. Acknowledging that all was not well with the state of affairs in Punjab agriculture, the state government in 1985 appointed a committee under the chairmanship of Professor S.S. Johl, an agronomist, to look into the problems of the agrarian sector.

In the report submitted in 1986, the Johl Committee expressed concern about stagnating productivity levels and deteriorating environment due to the cropping pattern dominated by paddy-wheat rotation. The committee recommended that if agriculture in Punjab was to be made sustainable, the farmers will have to be encouraged to diversify cropping pattern, switching over from high-volume and low-value crops to low-volume and high-value crops. However, given the overall political atmosphere in the state at that time, no concrete steps could be taken to implement the report.

Introduction of liberalization and globalization during the early 1990s further increased pressure on the agrarian economy. The 'new' economic policy advocated withdrawal of the state from economic sphere, leaving it to the logic of market forces. While it might be a good thing for the industry to be allowed to freely import the latest technology from abroad or have a competitive atmosphere, leaving the agricultural sector to the vagaries of free market could prove disastrous. Small landholders cultivate most of the land in

India and they often have to borrow from various sources for investments in the cultivation of cash crops. The cycle of agricultural production is such that virtually the entire farm yield comes to the market simultaneously. In a completely free and open market, the indebted small cultivator would obviously find it hard to bargain with the mighty trader.[5] The support price regime for food grain crops had been a great help to the farmers.

Notwithstanding the shift in economic policies, the agrarian lobby was able to prevail and the support price regime was not withdrawn. However, procurement agencies became lackluster and lethargic in procuring crops from farmers. With the extension of Green Revolution to other parts of India, the demand for food grains from Punjab also declined. Thus, even when the support price regime continued, the central government no longer raised the support prices much in the subsequent years.

This was most visible during the paddy procurement season of the year 2000. There had been a bumper crop of paddy in the state with no natural calamities like untimely rains or floods. But when the crop was brought to the *mandis* (marketing centres) the farmers were surprised to find that procurement agencies were not willing to buy their grains at the minimum support price declared by the central government. The officials claimed that they could not buy paddy because it was of inferior quality. The Food Corporation of India (FCI) chief went to the extent of saying that as much as 80 per cent of the Punjab paddy was spoilt—a claim that had no scientific basis. Indeed the FCI officials rarely conducted any tests while rejecting a particular lot of paddy even when they were provided with kits to carry out such tests.[6]

Interestingly, private traders and rice millers were quite willing to buy the same paddy but at a price much lower than the official support price, which would have hardly met the farmers' costs for production of the crop. In the given situation many farmers eventually sold their paddy to traders. The traders paid them Rs. 400–500 per quintal for the 'super fine' variety of paddy against the official support price of Rs. 550. For the common variety of paddy, traders paid them Rs. 350–400 per quintal against the official support price of Rs. 510.[7] Some traders reportedly sold the same paddy to official agencies at the minimum support price a few weeks later.

Many farmers, however, chose to wait for the official agencies with their grains in *mandis*, in some cases for over two weeks. Local newspapers during the month of October 2000 were splashed with pictures of paddy piled up in the *mandis* and farmers sleeping on them. 'The grain was everywhere' and *mandis* were overflowing with heaps of paddy. Paddy was being downloaded wherever the farmers could find room—on roads, in school grounds, in public parks.[8] The farmers were obviously depressed and angry, perhaps more depressed than angry! As a newspaper report states:

> Though the farmer's anger is coming to a boil, his attitude towards the government officials is, surprisingly, the very reverse. With folded hands, he pleads with them to lift his produce, at times virtually falling at their feet to grant him a 'remunerative' rate. A telling symbol of the vice-like grip that the market binds him in.[9]

The farmers were at the mercy of officials. 'It is blood and toil for six months and we cannot afford to annoy the officials. The money we earn during these days will provide for our family during the next six months as well as help us purchase fertilizers for the forthcoming wheat crop', a farmer in Khanna Mandi, Asia's biggest grain market, told Bajinder Pal Singh, a newspaper reporter. However, not all of them could wait or bear the humiliation. There were several reports on the front pages of local newspapers during the month of October 2000 of the small and marginal farmers taking the extreme step of committing suicide out of frustration and humiliation.

INDEBTEDNESS AND SUICIDES

Marginalization of agriculture has had many far-reaching implications for the farming population of the region and elsewhere. There has been a general stagnation of the agricultural sector over the last decade or so. The available analyses show that by the early 1990s, paddy and wheat had already reached peak levels of productivity in Punjab (Government of Punjab, 2004: 39).

According to the *Economic Survey of Punjab 2003–4*, the primary sector of Punjab economy registered a negative growth at the rate of minus 2.38 per cent over the preceding year (2002–3). The share of agriculture sector to the gross state domestic product has also declined significantly from 33.06 per cent in 1993–4 to 24.43 per cent in 2002–3 (ibid.: 45). More focused studies carried out by economists also show that over the last decade or so, cash expenditure on crop production has been steadily rising for different crops. The compound growth rate of cash expenditure between 1974–5 and 1991–2 was nearly 9 per cent for wheat and more than 11 per cent for rice, two of the main crops grown by Punjab farmers (Shergill, 1998: 3).

Much of this growth in expenditure has been a direct result of the increasing use of commercial inputs such as pesticides, fertilizers and seeds, by the farmers of Punjab. Since all these inputs are purchased with cash, farmers have to invariably invest a substantial amount of cash in every crop. Given that their own resources are limited, they invariably have to borrow, either directly from the market or via commission agents through whom they sell the yields of their farm. In some cases the situation has become so desperate that local farmers have put boards outside their villages stating 'the village is on sale'.[10]

Some recent studies carried out by economists in different parts of the state provide us with abundant evidence of growing economic hardships of the cultivators of Punjab. A study carried out by H.S. Shergill during the middle of 1990s found that as many as 86 per cent of the respondent farmers had to routinely borrow from various credit agencies for short-term investment on crops. Nearly 27 per cent of all farmers borrowed for capital investments in farm machinery.

In terms of dependence on borrowed money, the smaller landholders were clearly in a much weaker position than bigger landholders. Though the bigger farmers also frequently borrowed for short- and long-term investments on land, many of them also had savings. The average per acre outstanding debt of the small farmers worked out to be Rs. 3,396 as against Rs. 1,398 and Rs. 1,599 for the medium and big farmers, respectively. As discussed below, other more recent studies, have found the levels of indebtedness to be much higher.

It is interesting to note that despite official efforts at making institutional credit available to the cultivators, a significant proportion of short-term borrowings (61.31 per cent) by all categories of farmers were from the commission agents in grain markets, the *arhtiyas.* As many as 63.85 per cent farmers regularly borrowed from them. The Primary Cooperative Credit Societies came next from where 51.31 per cent of the farmers borrowed for their short-term credit needs. Only 8.85 per cent of the farmers borrowed from commercial banks for short-term investments in land (Shergill, 1998).

Another study by a team led by Sucha Singh Gill found that indebtedness of the surveyed farmers who had committed suicide in the Malwa sub-region of Punjab was in the range of Rs. 10,000 and Rs. 6.5 lakh and the average outstanding debt was Rs. 1.25 lakh per farmer household.[11]

More recently, Singh et al. (2005) in their study of six villages selected from different sub-regions of the state found that as many as 78.40 per cent of all the farm households in Punjab were under debt. Only those with large holdings were relatively free from debt. Average outstanding debt of the sample population worked out to be as high as Rs. 92,394 of which nearly 58 per cent had been borrowed from non-institutional sources. The extent and the nature of debt of course varied across different categories of farmers. While the absolute amount of outstanding debt with the marginal and small farmers was lesser (Rs. 17,465 and 43,598 respectively), the share of non-institutional debt in their outstanding debt was much higher (74 and 70 per cent, respectively).

This study also confirmed the overall domination of *arhtiyas* in the local credit market. *Arhtiyas* accounted for as much as 50.51 per cent of all the outstanding debt with their respondent farmers. Some other studies, such as by Gill (2000) and Kaur (2002) also found *arhtiyas* being the major source of credit for farmers of Punjab. There are obvious reasons for *arhtiyas* being so dominant in the local credit markets. The available institutional credit was simply not enough. *Arhtiyas* filled the gap for the availability of credit from institutional sources and the total demand for credit in rural Punjab (Gill 2004). *Arhtiyas*' credit also involved lesser paper work and other bureaucratic hassles.

Singh et al. (2005) in their study also looked at the purpose of borrowing. Money was borrowed for both 'productive' (41 per cent) and 'unproductive' (59 per cent) purposes. Though all categories of farmers borrowed heavily for 'social' needs, the share of 'unproductive' borrowings was highest amongst marginal farmers (71 per cent) and comparatively lesser amongst large farmers (48 per cent). Nearly half of the money borrowed for 'unproductive' purposes was spent on marriages and other social functions.

The cost of informal credit is almost always higher than that of institutional credit. The *arhtiyas* typically charge a monthly interest of 2–3 per cent. Further, as underlined by the classical literature on agrarian social structure in India (Bhardwaj, 1974; Bhaduri, 1984; Bardhan and Rudra, 1978), informal credit invariably comes with other demands and pressures, that is, the interlocking of credit with the product market. Though the context of contemporary Punjab agriculture is no longer semi-feudal in any sense of the term, the informal credit market is invariably tied to the product market. An indebted farmer has to not only compulsively produce for the market but has also to sell his produce to/through the *arhtiya* to whom he is indebted. Anita Gill's study found as many as 84 per cent of the sample households in Patiala district and 51 per cent in Amritsar district being interlinked borrowers (Gill, 2004: 3746). It may be useful to end this section with a quote from her study:

> On the whole . . . informal lenders have survived despite all proclaimed policy measures. Their guise has changed to a lender, whose principal activity is not moneylending. Rather, credit contracts are now interlinked with contracts in other markets. In the study area, it is the sale of crop (i.e., output) which is interlinked with credit and the *arhtiya* has emerged as the main informal lender. By shifting to a better collateral (crop, instead of land), these lenders have not only strengthened their bargaining power in their principal activity, but are also able to exploit the borrowers to the hilt by charging exorbitant rates of interest. And yet, borrowers are forced to turn to them because formal credit is not only inadequate, availing it is also a cumbersome process, involves ownership of land

explicitly or implicitly, and a sizeable class of cultivators cannot offer much land for loans. A lower rate of interest in formal market, then, is hardly any incentive. (ibid.: 3748)

As has happened in some other parts of India, Punjab too experienced an increase in cases of suicides by farmers and landless labourers during the decade of the 1990s. The available literature tends to point to a clear link between the increasing economic hardships of the rural people, particularly the smaller farmers and landless labourers, and the growing numbers of suicides.

According to an internal report prepared by the Department of Agriculture of the state government in 2004, though the phenomenon of farmers' suicides started some time during the 1980s, it witnessed a sudden increase during the 1990s. The situation became alarming when in one year, that is 1997, as many as 418 cases of suicides were reported from rural Punjab. The report identified a total of 2,116 cases of reported suicides since the mid-1980s, and recognized that 'these figures were only of reported acts and many must have gone unrecorded'. Further, it underlines some interesting facts about these suicides. They were mostly concentrated in the Malwa districts of Sangrur, Bathinda, Ferozepur, Mansa and Faridkot. More than 70 per cent of those who killed themselves were small/marginal farmers or landless labourers. In most cases agriculture was their only source of livelihood. A large majority of them were engaged in the cultivation of wheat and paddy crops (65 per cent) or wheat and cotton (20 per cent). More than 70 per cent of them were Jutts, and with the exception of one, they were all males, from a relatively younger age group.

The report also recognized that because of the declining growth rates, agriculture was no longer profitable and the cultivators had no alternative sources of employment available to them. This scenario led to rising debts. However, in some cases conspicuous consumption and drug addiction were also important factors that led to chronic indebtedness of those who committed suicide.

Another study conducted by the Institute of Development and Communication (IDC) also pointed to a sudden increase in cases of suicides in Punjab. The number of suicides committed in the state experienced a sharp rise from the year 1992–3. In 1992–3 suicides in Punjab increased by a staggering 51.97 per cent while the comparable figure for the country as a whole was only 5.11 per cent. In the years 1993–4 the increase was 14 per cent in Punjab as against 5.88 per cent for the country. While there was a decline in the reported cases of suicides at the all-India level, the state of Punjab once again reported an increase of 57 per cent in 1994–5. The IDC report also recognized the fact that there was always a possibility of underreporting of suicides (IDC, 1998).

The IDC study also pointed to the suicides being concentrated in certain pockets of the state. The most-suicide prone district in Punjab was Sangrur where as many as 22.39 per cent of all the suicides reported occurred during 1988–97. Another study reported that within Sangrur also, it was from certain blocks that most of the cases were reported. There were 12 specific villages in Lehragaga, Andana and Barnala blocks of Sangrur district where most of the suicides had occurred (Iyer and Manick, 2000).

Though the IDC study recognized the presence of a link between indebtedness and suicides, the explanations offered for the indebtedness of those who committed suicide is at variance from the official report and some other studies.

According to the IDC study, the contemporary crisis of Punjab agriculture emanated from: (a) limitations of the Green Revolution and lack of inner dynamism to build-up forward and backward inter-sectoral linkages; (b) decline in the size of operational holdings and fragmentation of land as well as pauperization of small and marginal farmers; (c) decline in the growth rate of productivity; and (d) increase in input costs and a corresponding fall in income of the small and marginal farmers.

Inferring from Shergill's study, the IDC study also pointed to the fact that small and marginal farmers had a much higher share of debt. However, it does not give much weight to indebtedness *per se* as being the main factor that explained the increase in suicides. Of the causes identified by the IDC study, indebtedness came at number three. Only 41.50 per cent of those who committed suicide were indebted. The corresponding figure for the 'general sample' was 71.40 per cent. The fact that 'only six per cent of the suicide victims had to sell land under the burden of debt' (IDC, 1998: 69) implies, according to the IDC study, that the debt burden was not particularly severe. The IDC study tends to suggest that 'social factors' were much more determining while explaining the rural suicides in contemporary Punjab. The most important factor was 'family discord' followed by 'alcohol and illicit drug use'. Even indebtedness in most cases was 'socially induced'. It may be worthwhile quoting from the study:

> . . . greater proportion of the debt of the small and marginal farmers originated from loans taken for non-productive purposes . . . a large amount of debt is socially induced among these sections. 68 per cent of the suicide victims' families have a debt on them because of unproductive expenditure as compared to 20 per cent of general households.
>
> . . . A number of suicides were noticed among those for whom the use of credit for conspicuous consumption had aggravated the stress situation. (ibid.: 36–7)

In contrast to the IDC study, Iyer and Manick (2000) treat the crisis of the agrarian economy and the growing indebtedness of farmers as the foremost cause of increase in suicides in rural Punjab. They make a crucial distinction between the 'causative' and 'precipitant' factors while explaining these suicides. While the precipitant factors could be social and psychological, the causative factors, in most cases, were economic (primarily indebtedness). According to them, the causative factors are the ones that produce those social conditions under which an individual begins to feel insecure and helpless.

In their sample, nearly 79 per cent of those who committed suicides came from poor families—mostly marginalized farmers or landless labourers. Only around 6 per cent of those who committed suicides were free from debt and most of them had borrowed money from informal sources, generally from the *arhtiyas*. In some cases, drug addiction and marital disputes also became causative factors, but these factors were not as critical for them as they were for the IDC study. Similarly, though indebtedness was a crucial and determining factor, by itself it could not be a sufficient cause for committing suicide. In most cases, it was the loss of honour and constant experience of humiliation at the hands of lenders that seemed to have 'precipitated' them to take such an extreme step.

EXPLAINING THE CRISIS OF AGRICULTURE

As mentioned above, one of the major implications of the shift in economic priorities during the early 1990s was a near complete marginalization of the rural society and agrarian economy in the popular discourses in India. However, the sudden spurt in suicides by farmers in different parts of the country has brought agriculture back into the public arena. Academics, activists, and policy makers are all talking about agriculture once again. However, this time agriculture is not being talked about in the manner in which it was during the early decades after Independence, when the farmer was presented as a food-giver, selfless and hard working, a symbol of national pride. Unfortunately, it is in the context of an unprecedented crisis of the agrarian sector, reflecting itself in seemingly unending reports of suicides by farmers, mostly from regions that had experienced agricultural development during the decades of 1960s and 1970s that agriculture is being talked about in India today.

Apart from public outcry on the subject, some scholars have also been analysing this phenomenon. Though overlapping significantly in their orientation and arguments, we can identify broadly different sets of perspectives in the current writings on the crisis of agriculture.

(i) First and foremost is 'the thesis of neglect'. According to this line of argument, the crisis of agriculture is a direct offshoot of the shift in priorities during the early 1990s. Growing obsession with the so-called 'new economy', information technology, media and the urban consumers led to a complete marginalization of the 'rural' and agrarian sector. The policies of economic liberalization, according to this perspective, also required the state to open up all sectors of the Indian economy to global market. Following this line of argument Chandrasekhar and Ghosh write:

> Public agricultural extension services have all but disappeared, leaving farmers to the mercy of private dealers of seed and other inputs such as fertiliser and pesticides who function without adequate regulation, creating problems of wrong crop choices, excessively high input prices, spurious inputs and extortion. Public crop marketing services have also declined in spread and scope, and

marketing margins imposed by private traders have therefore increased. All this happened over a period when farmers were actively encouraged to shift to cash crops, away from subsistence crops which involved less monetised inputs and could ensure at least consumption survival of peasant households. (Chandrasekhar and Ghosh, 2004)

This neglect of agriculture and rural economy has affected not only the farming communities but also the landless labourers who, because of the crisis in agriculture, are finding it much more difficult to find employment on viable wages. Commenting on this process, Jayati Ghosh writes:

The complete collapse of rural incomes or job opportunities has created an almost unprecedented situation of desperation among the landless, who rely exclusively on wage labour to survive. Some of this problem originates further back, in the inadequate development of non-agricultural work opportunities in most of rural India. This was directly related to the decline in public expenditure on rural development, which had adverse multiplier effects on rural non-agricultural economic activity in general. More recently, farm-related activities such as dairy have been hit by the decline of cooperatives (some of them killed by official design) and the pattern of trade liberalization. (Ghosh, 2004)

The recent realization by policy makers in the Government of India also seem to be working with a similar kind of understanding of the crisis of agriculture.

(ii) The second and perhaps more popular has been the thesis that links the rise in farmers' suicides to 'the crisis of ecology' and 'disintegration of community' caused by new technology and accentuated by globalization.

Commenting on the earlier crisis of militancy in Punjab, Vandana Shiva had, for example, argued that the Green Revolution was not merely a technological innovation meant for increasing productivity of land and bringing prosperity to farmers. Its negative social consequences far-exceeded its benefits. Green Revolution introduced a commercial culture in rural Punjab and destroyed the community. It changed social relations, from those based on mutual obligation to those based purely on the market principle. After the Green Revolution

Atomized and fragmented cultivators related directly to the state and the market. This generated on the one hand, an erosion of cultural norms and practices and on the other hand, it sowed the seeds of violence and conflict.

(Shiva, 1991: 171)

More recently, she along with some others attributed the suicides by cotton farmers directly to the ecological crisis generated by the introduction of new economic policies associated with the globalization process:

The tragedy of farmers committing suicides for a couple of years in some states, highlights some of these high social and ecological costs which are linked to globalisation of non-sustainable agriculture and which are not restricted to the cotton growing areas of various states but have been experienced in all commercially-grown and chemically-farmed crops in all regions. While the benefits of globalisation go to the seeds and chemical corporations through expanding markets, the cost and risks are exclusively borne by the small farmers and landless peasants. (Shiva et al., 1999)

Elsewhere Shiva writes:

. . . as farming is delinked from the earth, the soil, the biodiversity, and the climate, and linked to global corporations and global markets, and the generosity of the earth is replaced by the greed of corporations, the viability of small farmers and small farms is destroyed. Farmer suicides are the most tragic and dramatic symptom of the crisis of survival faced by Indian peasants. (Shiva, 2004).

Several other scholars have also tried to link the increase in farmers' suicides to a general breakdown of the ecological balance, disintegration of 'community' and kinship support system, and the rise of some kind of individualistic orientations brought in by new technology and development philosophy to the Indian countryside (Ahlawat, 2003; Vasavi, 1999).

(iv) Some have also attributed the agrarian distress leading to suicides in Punjab to the generic 'irrationality of the peasant'. The fact that in most cases the small and marginal farmers went into chronic indebtedness primarily because of 'unproductive' borrowings (as discussed earlier) implies that they had not 'acquired Weberian rationality characteristic of modern capitalism'. Quoting Thomas and Znaniecki, Singh (2005) finds basic irrational traits among Punjab cultivators. Instead of investing their returns from agriculture in further expanding their capital base, he argues, the Punjab farmers tend to spend everything on 'consumer goods to seek physical and sensual gratification'. A similar kind of conclusion could also be inferred from the IDC study of farmers' suicides in Punjab discussed above.

(v) There has also been another set of arguments that looks at the current crisis of Indian agriculture as having been entirely a consequence of the new economic policy and globalization, which, according to this position is basically a 'return to the colonial logic' of global economic integration. The most vocal advocate of this approach has been Utsa Patnaik. She draws parallel between what happened during the colonial period when the Indian peasants were compelled to produce commercial crops like cotton, because the British needed it for the textile mills in England, and the current phase of globalization. She argues:

> . . . lakhs of small farmers, were switching from food crops to cotton as the world prices were rising. Many of them had not cultivated cotton before. . . . There was this sudden expansion of area under cotton—these farmers could not afford to do so except on the basis of credit. They took loans and the amount of loans they took to produce cotton was much higher than they had taken in the past, as they would have grown rain-fed food crops on the same land, which would not have cost much for production. *So the switch to an exportable commercial crop led to a scenario of rising indebtedness* (emphasis added).
>
> (Patnaik, 2004)

She continues:

> . . . there is an interesting parallel that one can find with what happened during the cotton boom. In 1861, when the American Civil War broke out and supplies of raw cotton from the United States to the manufacturing centres in Britain and Europe were cut off, they turned to alternative sources of cotton and India as a major source. Suddenly the prices of global cotton went up and the Indian farmer, being always very price responsive, switched over from food crops to cash crops. Immediately, there was a huge expansion of areas growing cotton and a switch from food crops like *jowar* and *ragi* to cotton. In order to do so, they borrowed from the *sahukars* (moneylenders).
>
> When the Civil War ended, the global prices crashed. The story repeats itself in 1996. . . . But what happened was that when the people switched from food

crops to cash crops, the food prices went up. With the crash in cotton prices, the farmers found that they could not pay the *sahukars*. And the *sahukars* began to foreclose the debts. This led to the Deccan Riots. What happened then was that the Indian farmers actually took on the *sahukars* and unitedly fought against them, but this time they seem to be taking it out on themselves. (ibid.)

BEYOND CRISIS

The sudden spurt in cases of suicides by small and marginal farmers in different parts of India over the last couple of years has indeed created a sense of unprecedented crisis in Indian agriculture. The fact that this has happened simultaneously in different regions of India—from Karnataka and Andhra Pradesh to Maharashtra and Punjab—has understandably made social scientists, activists, and policy makers wonder about the causes and connection that such a phenomenon may have with wider processes of change operating at the national/global levels.

The above discussed explanations offered by different scholars are certainly persuasive. They make several useful suggestions about the prevailing state of affairs in Indian agriculture. However, when one looks at these arguments a little dispassionately, with a specific regional context in mind, one also begins to realize that the story of agrarian distress is perhaps more complex and complicated than what is made out to be in most of these 'theses'.

First and foremost, the notion of agriculture and rural social structure that they all seem to work with appears to be rather problematic. Even when vulnerability of small/marginal farmers is underlined in most of these analyses, they seem to treat agriculture in totality, as a sector of the wider Indian economy which has been pushed into crisis by various policy changes. The social context of agriculture, the village, is also seen as a generalized category. Such populist frames tend to look at village in communitarian terms, viz., peasant communities that were hitherto living undisturbed in peace and harmony, and have suddenly been pushed into crisis by commercial capital and global markets. Such analyses thus end up ignoring the internal differences of caste and class that are so obvious and important aspects of the Indian agrarian scene.

These analyses also tend to 'over-generalize' the nature of crisis and its causes, almost completely ignoring regional variations and diverse trajectories that mark the Indian agrarian scene today. Though globalization has indeed had many negative implications for the agrarian economy in general, the crisis of agriculture today is not being experienced in a similar way everywhere. A close look at contemporary Punjab, for example, clearly shows that suicides were largely localized in certain pockets and only a few cases were reported from most other parts. The broader context of agrarian change also varies a great deal.

EMERGING SCENARIO IN PUNJAB AGRICULTURE

It may be worthwhile to go back to the experience of agrarian change in contemporary Punjab in order to make sense of what is happening today. The impact of Green Revolution was not confined to village/agriculture alone. It transformed virtually everything—society, culture and politics. The economic development experienced during the Green Revolution period also brought the villages closer to the city life and its economy. S.S. Gill had reported this process some twenty years back:

> With the penetration of capitalist relations in agriculture, modern education has spread. Most of the Punjab villages have schools and some even have colleges functioning in them. Some of the capitalist farmers . . . are actually sending them to urban centres to acquire better education. With this a large number of educated persons from rural areas have been coming forward to take up jobs in government and semi-government institutions and departments. This has produced a distinct category of middle class intellectuals of rural origin. (Gill, 1985)

This process has become further pronounced since Gill's observation some twenty years back. Most agricultural households in contemporary Punjab have become economically diversified. As Lindberg rightly points out, they are increasingly becoming pluri-active, 'standing between farming and other activities whether as seasonal labourers or small-scale entrepreneurs in the local economy. . . . Agriculture and farming is no more an all-encompassing way of life and identity' (Lindberg, 2005: 11).

The available official data on employment patterns in Punjab has also begun to reflect this process very clearly. As shown in Table 15.1, the proportion of cultivators in the total number of main workers in Punjab declined from 42.56 in 1971 to 31.44 in 1991, and further to 22.60 by 2001. While the share of cultivators has been consistently falling, that of the agricultural labourers had been rising until the 1991 census. However, in the decade 1991 to 2001, even their proportion declined significantly, from 23.82 to 16.30. In other words, though nearly 70 per cent of Punjab's population still lives in rural areas, only around 39 per cent of the main workers in the state were directly employed in agriculture. The comparable figure for the country as a whole was still above 58 per cent.

Table 15.2, further corroborates the point that Punjab agriculture is undergoing some interesting shifts. As shown in the table, the total number of landholdings in the state declined from 13,75,392 in 1971 to 10,27,127 in 1981. This decline of more than 3 lakh holdings obviously suggests a sudden move away from agriculture. While this process experienced a reversal during the 1980s, perhaps because of the Khalistan movement, it again seems to have gained momentum during the post-1990 period.

Who is moving out of agriculture? If we look at Table 15.2 carefully, the decline is most clearly visible in the category of marginal landholdings (from 37.63 per cent in 1970–1 to mere 18.65 per cent in 1995–6), followed by small landholdings. There is a corres-

TABLE 15.1: PROPORTION OF WORKERS ENGAGED IN AGRICULTURE IN PUNJAB

Category	1971	1981	1991	2001
Cultivators	42.56	35.86	31.44	22.60
Agri. labourers	20.11	22.16	23.82	16.30
Total	62.67	58.02	55.26	38.90

Source: *Human Development Report 2004* (*Punjab*), p. 46; *Census of India*, 2001 (www.censusindia.net).

TABLE 15.2: DISTRIBUTION OF OPERATIONAL LANDHOLDINGS

Size-class (in hectares)	1970–1	1980–1	1990–1	1995–6
Marginal (0–1)	5,17,568 (37.63)	1,97,323 (19.21)	2,96,131 (26.50)	2,03,876 (18.65)
Small (1–2)	2,60,083 (18.91)	1,99,368 (19.41)	2,03,842 (18.24)	1,83,453 (16.78)
Medium (2–4)	2,81,103 (20.44)	2,87,423 (27.99)	2,88,788 (25.85)	3,20,340 (29.31)
Large (4–10)	2,47,755 (18.02)	2,69,072 (26.20)	2,61,481 (23.40)	3,05,792 (27.98)
Extra-large (10 +)	68,883 (5.00)	73,941 (7.19)	67,172 (6.01)	79,612 (7.28)
Total	13,75,392 (100)	10,27,127 (100)	11,17,414 (100)	10,93,073 (100)

Source: *Human Development Report 2004 (Punjab)*, p. 41.

pondent rise in the medium and large landholdings over this period of 25 years or so.

Though average size of the holdings is growing in Punjab, the bigger farmer is not necessarily becoming more rural. While marginal and small cultivators seem to be moving out of agriculture, the bigger farmer is moving out of the village (Table 15.3). The big farmers of Punjab invariably have a part of their families living in the town. Their children go to urban schools/colleges, and they invest their surplus in non-agricultural activities.

Not only has there been a fragmentation of farming classes, the rural social structure has also undergone a near complete transformation over the last three or four decades. My recent study of changing caste relations in rural Punjab clearly reflects this process. As I have argued elsewhere (Jodhka, 2002), commercialization and mechanization of agriculture on the one hand and introduction of democratic political process on the other have together transformed caste relations in rural Punjab quite fundamentally. Over the last twenty years or so large proportions of Dalits in Punjab

TABLE 15.3: GROWTH RATES IN DIFFERENT SECTORS OF PUNJAB ECONOMY

Time-period	Industry	Agriculture	Livestock
1966–7 to 1979–80	8.22	3.18	6.10
1980–1 to 1990–1	9.12	4.87	5.70
1991–2 to 1998–9	8.49	0.37	5.10

Source: *Human Development Report 2004* (*Punjab*).

have consciously dissociated themselves from their traditional occupations and have also been trying to distance them from everyday engagement with the agrarian economy, which were the sources of power for the locally dominant castes over them.

Their autonomization from the 'traditional' rural economy and structures of patronage and loyalty has created a rather volatile situation. While the institutions supporting ideas and structures of hierarchy have nearly disintegrated, the upper castes have not yet shed their prejudice against the former 'untouchable' groups. Nor have they yet reconciled to the changed ground realities. In the emerging scenario, local Dalits have begun to assert for equal rights and a share from the resources that belong commonly to the village and had so far been in the exclusive control of the locally dominant caste groups or individual households (Jodhka and Louis, 2003; Jodhka, 2004).

These processes of change have had a direct implication for the political agency of the farming classes in Punjab. The earlier solidarity of farmers reflected in their powerful mobilization during the 1980s is nowhere to be seen today. The Bharatiya Kisan Union (BKU), which had provided leadership to Punjab farmers during the 1980s, is split into four factions. Apart from the different factions of BKU, communist parties and other leftist groups also have their farmers' unions. Put together there are a total 10 different organizations claiming to represent farmers' interests in Punjab (Gill, 2004). Though some of them occasionally come together on common platform, most of the time they remain divided. This fragmentation of farmers' movement during the last decade or so cannot be understood without referring to the growing internal differentiation within the landowning classes/castes and the changing political and class balance at the local and regional level.

Further, the process of globalization/shift in economic priorities, and consequent weakening of the Indian State have all played their role in making the farmers' politics less effective. As in other parts of Asia, the Green Revolution was primarily a state-driven programme. Liberalization and the growing emphasis on withdrawal of State and the increasing involvement of the corporate sector in agriculture has changed the opportunity structure of farmers' politics (Lindberg, 2005). The moral high ground of Punjab's farmers as producers of food for the nation is lost when they are told to diversify into non-food grain crops that can find markets, locally and globally, without any help from the State.

Notwithstanding all these new trends, the change on ground is still limited. Farmers in Punjab, as elsewhere in India, continue to be an influential block. Populist politics often resorted to by the regional elite works to their advantage. This is quite evident from the available data on agricultural subsidies and the state policies during the last more than a decade of liberalization/globalization. In the case of Punjab, for example, after facing problems with procurement for about two years, crops were being procured again by the state agencies and the support prices were rising, albeit slowly. Instances of farmers committing suicides in rural Punjab have come down in the last couple of years as compared to the decade of 1990s.

However, this is not to suggest that the crisis of Punjab agriculture is over. On the contrary, as is reflected in the literature on growing indebtedness, the challenges before the farmers of Punjab seem to be growing. The fact that smaller landholders are increasingly finding it hard to stay in agriculture and are moving to other occupations clearly shows the nature of pressures the agriculturists in Punjab, as elsewhere in India, are confronted with. However, one should also see it in relation to the larger changes being experienced at the village and regional levels—economic, social and cultural. This perhaps would enable us to raise more meaningful questions about what is happening to agriculture today.

NOTES

1. *The Tribune*, Chandigarh, 21 March 2001.
2. This move was predictably criticized by various people for different reasons, one of them being that such a provision could in fact encourage those in distress to take their life, amounting to abetment of suicide. Some described the move as being communal, suggesting that the Akali government wanted to appease the farmers because most of them were Sikhs and the party's 'vote bank'. See, for example, 'Will More Farmers Commit Suicide Now?' by Chander Prakash, *The Tribune*, Chandigarh, 21 March 2001.
3. As in the official website of the Punjab Government, 2004 <punjabgovt.nic.in>.
4. Income of the secondary sector of the Punjab economy during the 1980s grew at the rate of 7 per cent per annum as against the all India average of 6.1 per cent (Gill and Ghuman, 2000: 450).
5. There have been several reports on the growing indebtedness of the Punjab farmers, particularly those of small and marginal farmers. Their sources for borrowings are invariably informal, generally the *arhtiyas* (the commission agents) in the grain market, which obviously makes their bargaining position weak. See Shergill, 1998; Bose, 2000.
6. H. Jaisingh, 'Growing Frustration of Kisan: Agriculture Needs Fresh Strategy', *The Tribune*, Chandigarh, 11 October 2000.
7. As reported by Lalit Mohan in *The Tribune*, 10 October 2000.
8. As reported by Bajinder Pal Singh, *The Indian Express*, 5 October 2000.
9. Ibid., 6 October 2000.
10. These villages are Harkishanpura in Bathinda district and Shergarh in Muktsar district. As in S.S. Gill, 2004.
11. *The Tribune*, 21 October 2000.

REFERENCES

Ahlawat, S.R. (2003), 'Sociology of Agrarian Crises: Peasant Suicide and Emerging Challenges', *Man and Development*, vol. XXV (3), pp. 97–110.

Banaji, Jairus (1996), 'The Farmers' Movement: A Critique of Conservative Rural Coalitions', *Journal of Peasant Studies*, vol. 21 (3 and 4), pp. 228–45.

Bardhan, P.K. and Ashok Rudra (1978), 'Interlinkage of Land, Labour and Credit Relations: An Analysis of Survey Data in East India', *Economic and Political Weekly*, vol. XIII (nos. 6 and 7), Annual Number, pp. 367–84.

Bhaduri, A. (1984), *The Economic Structure of Backward Agriculture*, New Delhi: Macmillan.

Bharadwaj, K. (1974), *Production Conditions in Indian Agriculture*, Cambridge: Cambridge University Press.

Bose, A. (2000), 'From Population to Pests in Punjab: American Boll Worm and Suicides in Cotton Belt', *Economic and Political Weekly*, vol. XXXV (38), pp. 3375–7.

Chandrasekhar, C.P. and Jayati Ghosh (2004), 'Agrarian Crisis in Andhra Pradesh', *Business Line*, 5 October.

Das, V. (ed.) (1990), 'Introduction', *Mirrors of Violence: Communities, Riots and Survivors in South Asia*, New Delhi: Oxford University Press.

Ghosh, Jayati (2004), 'The Rural Gloom', *Frontline*, vol. 21, issue 24, 20 November–3 December.

Gill, Anita (2004), 'Interlinked Agrarian Credit Markets: Case Study of Punjab', *Economic and Political Weekly*, vol. XXXIX (33), pp. 3741–51.

——— (2000), *Rural Credit Markets: Financial Sector Reforms and the Informal Lenders*, New Delhi: Deep and Deep.

Gill, S.S. (2004), 'Punjab Farmers' Movement: Continuity and Change', *Economic and Political Weekly*, vol. XXXIX (27), pp. 2964–6.

——— (1985), 'Genesis of Punjab Problem', in Abida Samiuddin (ed.), *The Punjab Crisis: Challenge and Response*, Delhi: Mittal Publishers.

Gill, S.S. and R.S. Ghuman (2000), 'Crisis of Punjab Economy: The Alternative Options and the Role of the Government', in R.S. Bawa and P.S. Raikhy (eds.), *Punjab Economy: Emerging Issues*, Amritsar: Guru Nanak Dev University, pp. 437–58.

Government of Punjab (2004), 'Suicides by Farmers in Punjab', Chandigarh, (unpublished official note).

——— (2000), *Statistical Abstract of Punjab*, Chandigarh: Punjab Government.

Gupta, D. (1996), *The Context of Ethnicity: Sikh Identity in a Comparative Perspective*, New Delhi: Oxford University Press.

Human Development Report 2004: Punjab, Chandigarh: Government of Punjab.

Institute for Development and Communication (IDC) (1998), *Suicides in Rural Punjab*, Chandigarh: Institute for Development and Communication.

Iyer, K.G. and M.S. Manick (2000), *Indebtedness, Impoverishment and Suicides in Rural Punjab*, Delhi: Indian Publishers & Distributors.

Jodhka, S.S. (2004), 'Sikhism and the Caste Question: Dalits and Their Politics in Contemporary Punjab', *Contributions to Indian Sociology* (n.s.), vol. 23 (1&2), pp. 165–92.

——— (2002), 'Caste and Untouchability in Rural Punjab', *Economic and Political Weekly*, vol. 37 (19), pp. 1813–23.

——— (ed.) (2001b), *Community and Identities: Contemporary Discourses on Culture and Politics in India*, New Delhi: Sage.

——— (2001a), 'Looking Back at the Khalistan Movement: Some Recent Researches on its Rise and Decline', (Review Article), *Economic and Political Weekly*, vol. XXXVI (16), pp. 1311–18.

——— (1997), 'Crisis of the 1980s and Changing Agenda of 'Punjab Studies': A Survey of Some Recent Research' (Review Article), *Economic and Political Weekly*, vol. XXXII(6), pp. 273–9.

Jodhka S.S. and P. Louis (2003), 'Caste Tensions in Punjab: Talhan and Beyond' (with Prakash Louis), *Economic and Political Weekly*, vol. XXXVIII (28), pp. 2923–6.

Kaur, Ravinder (1986), 'Jat Sikhs: A Question of Identity', *Contributions to Indian Sociology*, vol. 20 (2), pp. 221–40.

Kaur, Gian (2002), 'Role of Informal Rural Financial Markets in Punjab: A Case Study', *Working Paper 1*, NABARD, Mumbai: Department of Economic Analysis and Research.

Kohli, D.S. and N. Singh (1997), 'The Green Revolution in Punjab, India: The Economics of Technological Change', Conference paper.

Lindberg, Steffan (2005), 'Whom and What to Fight?: Indian Farmers Collective Action under Liberalisation and Globalisation', See Chapter 5 in this volume.

Patnaik, Utsa (2004), 'It is a Crisis Rooted in Economic Reforms', *Frontline* (net edition), vol. 21(13), 19 June–2 July.

Pettigrew, J. (1995), *The Sikhs of the Punjab: Unheard Voices of State and Guerrilla Violence*, London: Zed Books.

——— (1992), 'The Jats of Punjab' in Dipankar Gupta (ed.), *Social Stratification*, New Delhi: Oxford University Press.

Saberwal, S. (1987), *India: The Roots of Crisis*, New Delhi: Oxford University Press.

Shergill, H.S. (1998), *Rural Credit and Indebtedness in Punjab*, Chandigarh: Institute for Development and Communication (monograph series IV).

Shiva, Vandana (2004), 'The Suicide Economy of Corporate Globalisation', 5 April, *Znet* (www.zmag.org).

——— (1991), *The Violence of the Green Revolution: Ecological Degradation and Political Violence in Punjab*, London: Zed Books.

Shiva, Vandana, Ashok Emani and Afsar H. Jafri (1999). 'Globalisation and Threat to Seed Security: Case of Transgenic Cotton Trials in India', *Economic and Political Weekly*, vol. XXXIV (10), pp. 601–13.

Singh, B.P. (2005), 'Punjab Peasantry in Turmoil', see 'Introduction' to this volume.

Singh, Sukhpal et al. (2005), 'Magnitude and Determinants of Indebtedness in Punjab Agriculture', see Chapter 6 in this volume.

Vasavi, A.R. (1999), 'Agrarian Distress in Bidar: Market, State and Suicides', *Economic and Political Weekly*, vol. XXXIV (32), pp. 2263–8.

World Bank (2004), *Resuming Punjab's Prosperity: Opportunities and Challenges Ahead*, New Delhi: Poverty Reduction and Economic Management Sector Unit, South Asia Region.

Contributors

AIJAZ AHMAD: Formerly Fellow of Nehru Memorial Museum & Library, Delhi and Professor, Khan Abdul Ghaffar Khan Chair, Jamia Millia Islamia, Delhi.

AASIM SAJJAD AKHTAR: scholar and social activist, Islamabad.

BALKAR SINGH DAKAUNDA: Secretary, Bharatiya Kisan Union (Ekta), Punjab.

SUCHA SINGH GILL: Professor, Department of Economics, Punjabi University, Patiala.

SURINDER SINGH JODHKA: Professor, Centre for the Study of Social Systems, Jawaharlal Nehru University, Delhi; Director, Institute of Dalit Studies, Delhi.

STAFFAN LINDBERG: Professor, Department of Sociology and Director SASNET, Lund University, Lund, Sweden.

UTSA PATNAIK: Professor, Centre for Economic Studies and Planning, Jawaharlal Nehru University, Delhi.

RONKI RAM: Reader, Department of Political Science, Panjab University, Chandigarh.

ZIA UR REHMAN: Research Associate, Sustainable Development and Planning Institute, Islamabad.

LUBNA SAIF: Professor and Chairperson, Department of Pakistan Studies, Allama Iqbal Open University, Islamabad.

AHMAD SALIM: Director, Urdu Publications, Sustainable Development and Planning Institute, Islamabad.

HARJINDER SINGH SHERGILL: Professor, Department of Economics, Panjab University, Chandigarh.

PASHAURA SINGH SIDHUPUR: President, BKU (Ekta), Punjab.

BIRINDER PAL SINGH: Professor, Department of Sociology & Social Anthropology, Punjabi University, Patiala and former Fellow, Indian Institute of Advanced Study, Shimla.

SUKHPAL SINGH: Professor, Centre for Management in Agriculture, Indian Institute of Management, Ahmedabad.

DARSHAN SINGH TATLA: Fellow, Department of Theology, Birmingham University, Birmingham (UK) and Director, Punjab Centre for Migration Studies, Jalandhar.

Index